Six Chemicals That Changed Agriculture

Six Chemicals That Changed Agriculture

Robert L. Zimdahl
Professor Emeritus, Colorado State University, Fort Collins,
Colorado, USA

AMSTERDAM • BOSTON • HEIDELBERG • LONDON
NEW YORK • OXFORD • PARIS • SAN DIEGO
SAN FRANCISCO • SINGAPORE • SYDNEY • TOKYO

Academic Press is an imprint of Elsevier

Academic Press is an imprint of Elsevier
125 London Wall, London EC2Y 5AS, UK
525 B Street, Suite 1800, San Diego, CA 92101-4495, USA
225 Wyman Street, Waltham, MA 02451, USA
The Boulevard, Langford Lane, Kidlington, Oxford OX5 1GB, UK

Library of Congress Cataloging-in-Publication Data
A catalog record for this book is available from the Library of Congress

British Library Cataloging-in-Publication Data
A catalog record for this book is available from the British Library

ISBN: 978-0-12-800561-3

For information on all Academic Press publications
visit our website at http://store.elsevier.com/

Working together
to grow libraries in
developing countries

www.elsevier.com • www.bookaid.org

Publisher: Nikki Levy
Senior Acquisition Editor: Nancy Maragioglio
Editorial Project Manager: Billie Jean Fernandez
Production Project Manager: Chris Wortley
Designer: Mark Rogers

Typeset by TNQ Books and Journals
www.tnq.co.in

Printed and bound in the United States of America

Contents

10. Conclusion

Epigraph

We are "the result of our own past actions and our own
present doings. We ourselves are responsible for our
own happiness and misery. We create our own heaven.
We create our own hell. We are the architects of our own fate."[1]

---

1. This is a general statement about karma, typical of Hindu, and Buddhist thought. The general idea stems from the ancient Sanskrit tradition and is a tenet of much of South Asian thought. (I thank my colleague and friend Dr. James Boyd, Professor Emeritus, Philosophy, Colorado State University for clarifying this thought.)

Acknowledgments

The idea for this book was conceived by and the book would not have been completed without the editorial guidance, tolerance, and friendship of Dr. Rodney K. Skogerboe, Professor Emeritus of Chemistry, Colorado State University.

I acknowledge and appreciate the editorial guidance of Dr. Roland J. Buresh, Senior Scientist, International Rice Research Institute, Metro Manila, Philippines; Dr. Wilhelm F. Knaus, University Professor, Universität für Bodenkultur. Institut für Nutztierwissenschaften (University of Natural Resources and Life Sciences) Vienna, Austria; Dr. Jennifer Mclean, Associate Professor, Department of Microbiology, Immunology and Pathology, Colorado State University; Dr. Kenneth Watkins, Assistant Professor Emeritus of Chemistry, Colorado State University; Mr. Keith Pardieck, Patuxent Wildlife Research Center, US Geological Survey, Laurel, MD; Mr. Geoff LeBaron, Audubon Society, Washington, D.C.; and Mr. John Yule, Fort Collins, CO.

I also must acknowledge others who have contributed so much to me for many years. Their counsel, advice, and friendship have been essential to my writing, indeed to my scholarly/intellectual development. Dr. James W. Boyd, Professor Emeritus of Philosophy, Colorado State University, and Reverend Dr. Robert A. Geller; Fort Collins, CO. My wives Pamela J. Zimdahl (died 2012) and Karen R. Zimdahl, who encouraged my writing and offered comments and criticism when they thought they were appropriate, which they usually were.

Finally, errors of fact or interpretation are mine. I will, of course, appreciate comments about them from those who pay me the compliment of reading all or part of the book.

Chapter 1

Introduction

We live in a society exquisitely dependent on science and technology, in which hardly anyone knows anything about science and technology

Carl Sagan (Head, 2006)

CHEMISTRY

Chemistry is a physical science that studies the composition, structure, properties, changes, and reactions of elementary forms of matter. It deals specifically with atomic and molecular systems and seeks to know what a substance is composed of, how its properties are related to its composition, and how and why one chemical may or may not react with another. Chemists are also concerned with how atoms and molecules are involved in various forms of energy and energy production. Chemists study the concepts of dynamics, energetics, structure,

Six Chemicals That Changed Agriculture. http://dx.doi.org/10.1016/B978-0-12-800561-3.00001-8
1

composition, and synthesis. Some chemical reactions produce energy; others require it. Some reactions cause changes in phases of matter, separation of mixtures, and create the properties of polymers and other complex molecules. In chemistry, as in many scientific endeavors, things are not always exactly as they seem and unexpected, sometimes serendipitous things occur.

For many, chemistry may be remembered as the subject that was most difficult in school. Difficult it may be, but chemistry dominates our lives and agriculture in many ways, most of which few of us know much about. Chemistry is involved in, explains the details of, and helps us understand the food we eat, the air we breathe, the water we drink and bathe in, the clothes we wear, our emotions, and literally every object we see or touch every day. It is the interface between and the bridge that connects the subatomic world of the physicist and the molecular world of the biologist. Chemistry has improved our lives, caused some problems, and affected, frequently has determined, how our society and our agriculture have evolved and flourished.

If we are to understand chemistry's multiple positive and negative effects on our society and on how agriculture is practiced, we must know and learn from chemical and agricultural history. There are obstacles. In his 1959 Collected Essays, Aldous Huxley (1894–1963) admonished us "That men do not learn very much from the lessons of history is the most important lesson that history has to teach." Henry Ford (1916) in apparent neglect of the necessity of knowing the history, the story, said "History is more or less bunk." Ford actually said it, but seemed to have had a very shallow understanding of history and a lack of trust in the little he did understand (Lockerby, 2011). Henry David Thoreau, the careful observer of society told us, but not Mr. Ford, that there is always a story. "Wherever men have lived there is a story to be told, and it depends chiefly on the story-teller or historian whether that is interesting or not. You are simply a witness on the stand...."

The history we know is stories of the exploits of men and women that have caused change, development, or catastrophe. We tend to believe that historical events have occurred because of the direct intervention of people and for much of agricultural history that is true. People have determined the course of agricultural history. It is not all bunk, and the stories that follow verify we have learned from it. Knowing and learning from the stories of chemistry and chemists, the subjects of this book, are essential to understanding the development of the modern agricultural systems that feed the world.

ALCHEMY

Alchemy, a medieval science, began in pre-Pharaonic Egypt (before 3000 BC), flourished during the Hellenistic period (approximately 300 BC to the emergence of Rome 30 BC), and in twelfth-century medieval Europe (the Middle Ages: fifth to fifteenth centuries). It has remained as an interesting, influential philosophical tradition whose practitioners claimed it to be the precursor to profound powers[1]; that were never achieved, but it has had a direct influence on modern agriculture. Historically its defining objectives included three goals:

1. Creation of the fabled philosopher's stone; the ability to transform base metals into the noble metals (gold and silver).
2. Discovery of a universal cure for disease.
3. Development of an elixir of life, which would confer youth and longevity.

Alchemy is properly regarded as a protoscience, and, as such, it did not directly influence agriculture. It is properly regarded as a necessary precursor that contributed to the development and profound powers of modern medicine and chemistry; the latter has played a key role in development of modern agriculture. Alchemists developed, as scientists do, a framework of theory, terminology, experimental process, and basic laboratory techniques that is still recognizable. But its inclusion of Hermetic principles and practices related to mythology, magic, religion, and spirituality,[2] although interesting, distinguish it from modern chemistry. The alchemist's major goal, which they never accomplished, was to convert base metals (Al, Cu, Pb, Ni, Sn, Zn), which oxidize and corrode on exposure to air, and water, into the noble metals silver (Ag) and gold (Au).

Although regarded as a protoscience and in some circles as a pseudoscience, alchemy produced a wide range of contributions to medicine and the physical sciences. Robert Boyle (1627–1691), an alchemist, discovered the inverse relationship between the volume of a gas and pressure, known as Boyle's law. He also discovered the necessity of air for combustion and animal breathing. His work laid some of the foundation of modern chemistry. He was a pioneer of modern experimental scientific methods. Paracelsian[3] iatrochemistry emphasized the study of chemistry in relation to the physiology, pathology, and treatment of disease. Although we don't know how it is reported that alchemy influenced Isaac Newton's theory of gravity. Historical research supports the claim that alchemists searched for a material substance using physical methods.

Alchemists contributed to the chemical "industries of the day—ore testing and refining, metalworking, production of gunpowder, ink, dyes, paints, cosmetics, leather tanning, ceramics, glass manufacture, preparation of extracts, and liquors. Alchemists contributed alcohol distillation to Western Europe. Their attempts to arrange information on substances and to clarify and anticipate the products of their chemical reactions resulted in early conceptions of chemical elements and the first rudimentary periodic tables. They learned how

---

1. See http://en.wikipedia.orgm/wiki/Alchemy. Accessed September 16, 2013.
2. Hermetic=impervious to outside interference or influence. The seven Hermetic Principles are mentalism, correspondence, vibration, polarity, rhythm, cause and effect, and gender.
3. Auroleus Phillipus Theophrasus Bombastus von Hohenheim, Paracelsus (1493–1541), a Swiss physician and alchemist. Iatro is Greek for doctor.

to extract metals from ores, and how to compose many types of inorganic acids and bases."[4] Subsequently chemistry was rather dormant for several centuries before the development of modern chemical science. It is generally agreed that Antoine Lavoisier (1743–1794), a French chemist, should be regarded as the father of modern chemistry. His work was revolutionary because of his efforts to fit all experiments into the framework of a single theory and his invention of a system of chemical nomenclature that is still in use. He denied the existence of phlogiston—a hypothetical, volatile constituent of all combustible materials, which was released in flame. He wrote "The Elements of Chemistry" in1787.[5] Lavoisier's book was preceded by Boyle's 1661 book, "The Skeptical Chymist," widely regarded as a cornerstone book of chemistry.

Organic chemistry began when Friedrich Wöhler disproved that the theory of vitalism developed by Jöns Berzelius, a Swedish chemist, was false. Vitalism held that there were two categories of chemicals: organic and inorganic. Only the tissues of living creatures had the capability to produce organic chemicals. Therefore, it was not possible to synthesize an organic chemical from inorganic reactants. Wöhler, without specific intent, heated lead cyanate and ammonia and produced crystals of urea, an organic compound from inorganic reactants (see Chapter 4).

His work illustrates an important, essential tenet of science. A scientific hypothesis is falsifiable. New evidence, new data prove the previous theory was wrong.

CHEMISTRY AND THE CHEMICALS IN THIS BOOK

Modern chemists discovered, taught us how to use, and, in some cases, created the chemicals in this book. The purpose of this book is to discuss the history, some of the people involved, the synthesis, chemistry, and uses of some of the chemicals that have determined the structure, practice, and success of modern agriculture. The emphasis is on the chemicals that transformed agriculture from a system of intensive labor with limited to no predictability of crop or animal results, low yields, complete dependence on unpredictable weather, and limited to no external inputs into a system that needs less labor, has great predictability, requires regular external inputs, and has consistently, almost predictable, high yields.

The periodic table of the chemical elements arranges 128 elements by atomic number in such a way that their periodic properties are clear. The table includes 14 naturally occurring lanthanides (rare earth elements[6]), which have three valence electrons and 15 radioactive actinides—metallic elements with atomic numbers from 89 to 10 that have four valence electrons. Six actinides (actinium through

---

4. See http://en.wikipedia.org/wiki/Alchemy#Relation_to_the_science_of_chemistry.
5. http://chemistry.about.com/od/historyofchemistry/a/Who-Is-The-Father-Of-Chemistry.htm. Accessed January 2014.
6. The lanthanides are poorly named rare earth elements. They are not rare. They are relatively abundant. They tend to be fluorescent or phosphorescent. A common use provides color on television screens.

plutonium) can be found in nature; nine (americium through lawrencium) are man-made. The standard form of the table includes 7 periods (rows) and 18 groups (columns). Elements in groups have similar properties because they have the same number of valence electrons. The seven periods signify the highest unexcited energy level for an electron in that element. The evolution of the periodic table, first published by D. Mendeleev in 1869, is one of chemistry's major achievements.

More than 7 million chemical compounds have been made from two or more elements. From this multitude of chemicals a few have been selected for this book because of their widespread use and their transforming effects on agriculture. In full recognition that others could have been chosen, this book focuses on the agricultural contribution of three naturally occurring chemicals.

Lime: a soil amendment
Nitrogen: a fertilizer
Phosphorus: a fertilizer

Two synthetic organic chemicals.

2,4-D: an herbicide
DDT: an insecticide

and two chemical discoveries

Recombinant DNA: the necessary prelude to genetic engineering of crops
Antibiotics: with emphasis on their role in enabling confined animal feeding operations and their possible effects on human health

Each of these has changed the way agriculture is practiced and has made it possible to feed most of an expanding human population.

FIVE CHARACTERISTICS OF DEVELOPED COUNTRY AGRICULTURE

Five characteristics of today's agriculture in the world's developed countries, particularly in the United States have become apparent over the past six decades. They dominate agriculture in much of the world and distinguish modern agriculture from the practice of agricultural before 1950. The characteristics are:

1. Steadily increasing productivity
2. A continuing decline in the number of mid-sized farms
3. Increasing farm size
4. Increasing percentage of production by fewer, larger farms
5. Increasing dependence on technology

Steadily Increasing Productivity

There has been a sustained increase in agricultural production since the late 1930s, which was clearly evident by 1940. Since the Rev. T. R. Malthus (1766–1834) wrote his best seller "An Essay on the Principle of Population" in 1798 (he was 32), when the Earth may have had a billion people, there has been sustained doubt about the ability of food production to keep pace with human population growth. Increasing production has relieved but not eliminated doubt about agriculture's continued ability to feed the 9–11 billion people who may share the planet with us sometime between 2040 and 2050. Malthus' theory has been resurrected regularly since 1798. The Malthusian skeptics delight in reporting how consistently wrong he has been. Kolbert (2013) noted that every time "we're at the end of our proverbial rope...we find new rope." If human numbers increase geometrically as Malthus proposed, so, too ... does human ingenuity (Kolbert). The products of human ingenuity that have been so important to agriculture have nearly always originated in the chemist's laboratory. The Malthusians readily retort that Malthus will be right, but "his timing was off" (Kolbert). Advances in the chemical technology, which have been essential to ever-increasing agricultural productivity, will ultimately be insufficient because of an increasing human population on a finite earth.

Doubt about agriculture's ability to continually increase food production persists in spite of the fact that in my lifetime total wheat production in the United States has increased 100%, corn 400%, and soybeans, a crop almost unknown in the United States before 1940, a dazzling 1000%. In 2009, agricultural production was 170% above the 1948 level.[7] Increasing production yielded significant economic gains for farmers and for consumers who gained through lower food prices (Gardner, 2002. p. 261).

Lee Iacocca did not gain his fame and fortune in agriculture. However, he provided an insightful comment on the role of government policy in agriculture, which has been an essential part of agriculture's productive success. He said: (Iacocca and Novak, 1984, pp. 331–332).

> *It is unlikely that anyone of the causal factors that have been considered*
> *can qualify as* the *explanation for the takeoff and productivity, or for the*
> *longer-term history of technological change in U.S. agriculture......there's*
> *more going on here than good climate, rich soil, and hard-working farmers.*
> *We had all those things 50 years ago, and all we got were dust bowls and disasters.*
> *The difference lies in a wide range of government-sponsored projects. There are*
> *federal research projects; county agents to educate people; state experimental*
> *arms; rural education and irrigation projects......crop insurance; export credits;*
> *price support; acreage controls. With all that government help (or, some would*
> *say interference) we've created a miracle. Our agricultural/industrial policy*
> *has made us the envy of the world.*

Iacocca was right: modern agriculture has been favored by government policies that encourage chemical, capital, and energy-intensive agricultural practices. Government policy favors use of technology that increases

---

7. See www.theatlantic.com/magazine/archive/2012/07/thetriumph-of-the-family-farm/308998/#. Accessed August 2012.

production as opposed to a policy that favors integration of production goals with simultaneous achievement of social justice and environmental quality (Zimdahl, 2012, p. 132).

Modern, highly productive, industrial agriculture has been enabled by full employment of science and technology and, of great importance, externalization[8] of environmental and social costs. The real costs of environmental pollution, water contamination, nontarget species harm, and harm to human health are not borne by manufacturers, those who approve use, those who recommend, or the ultimate users of agricultural technology. The real costs are externalized and borne by society when food is purchased and taxes paid. However, agriculture's great productive achievements have not improved food security for those in the lower economic strata of society. Lower prices at the grocery store have not benefitted the percentage of Americans people in poverty, which in 2012 remained unchanged, as it has for several years, at 15%.

A Continuing Decline in the Number of Mid-Sized Farms

Beginning in 1840, US Bureau of the census data show that the US population, the number of farmers, the number of farms, and average acres per farm steadily increased. US population continued to grow, but the number of farms and farmers began to decline in the 1930s. Their number as a percent of the US labor force steadily decreased from 60% in 1860 to less than 1% in 2010. Ergo, 99% of Americans who are not farmers are absolutely dependent on those who are. The number of small farms (60% of all farms) with annual sales of less than $10,000 increased 1% from 2000 to 2007. The number of farms with sales greater than $10,000 but less than $100,000 decreased. Mid-sized farms with sales between $100,000 and $250,000 increased 7%. From 2000 to 2007, 11 states had a 4% increase in the number of farms, whereas 39 states decreased. The number of mid-sized farms has been decreasing since 1980 while the number of large farms and very small farms has gone up. The number of farms with sales over $250,000 grew from 85,000 to 152,000 from 1982 to 2002. Most growth came from farms with sales over $500,000.

Increasing Farm Size

The number of acres per farm continued to increase until 1990 when small farms, which had not been, were included in the US Department of Agriculture survey. Farm size peaked in 1990 at 461 acres and declined to 418 acres in 2010. Increasing farm size and the higher production of larger farms is consistent with

---

8. An externality is a cost that is not reflected in price. It is a cost or benefit for which no market mechanism exists. In the accounting sense, it is a cost that a decision-maker does not have to bear, or a benefit that cannot be captured. From a self-interested view, an externality is a secondary cost or benefit that does not affect the decision-maker. It can also be viewed as a good or service whose price does not reflect the true social cost of its consumption.

agriculture's dominant ethic—productionism. As long as agriculture continues to produce more it is doing the right thing, the ethical thing, that which should not be questioned. It is the view supported by commercial farmers, agribusiness, and the agricultural research/technology establishment centered in land-grant universities. It emphasizes the virtue of technical progress, low cost food, and the efficiency of large-scale commercial/industrial agriculture (Gardner, 2002, p. 351). The alternative view emphasizes ecological sustainability, small farms, and skepticism about the ultimate sustainability of an agricultural system that is capital, chemical, and energy dependent.

The small, family farm is widely respected, perhaps revered for the values it is presumed to represent. Family farms are under economic pressure, but they continue to demonstrate great resilience. Family farms are not necessarily small. Many of the largest farms are family farms, because families have formed corporations to take advantage of legal and accounting benefits. The most common criticism directed at those who advocate maintenance of small family farms is "if we are going to feed the world we can't go back to 40 acres and a mule."

The economic, indeed survival, pressure on farming and farmers has dictated that many have obeyed the counsel of the former Dean of Agriculture at Purdue University and US. Sec. of Agriculture (1971–1976) in the Nixon and Ford administrations, Earl Butz—"get big or get out," "plant fence row to fence row." The gradual demise, perhaps ultimately the disappearance, of the family farm is clear. Wendell Berry,[9] author of more than 50 books, whose dominant focus is agriculture, is one of the few persistent spokesman for the value of the small, self-supporting, sustainable family farm. Berry personifies an American school of thought that was notable, but also contested, in the founding generation. The debate set Thomas Jefferson against Alexander Hamilton, rural farms against cities, and agriculture against banking interests. Berry stands with Jefferson. He stands for local culture and the small family farmer, for yeoman virtues and an economic and political order that is modest enough for its actions and rationales to be discernible. Government should recognize and applaud the ground beneath our feet (the soil life depends on) and our connections with our fellow human beings. Its solutions should be equal to its problems and should not beget other problems. While the data are clear, neither public nor legislative debate occurs. The inexorable trend is allowed to continue.

Biodiversity, the primary focus of an Economist (2013) magazine special report, questions the validity of the view of agriculture that claims to have achieved sustainability (a universally accepted, good goal). The authors recommend that those who practice agriculture must recognize that ecological sustainability is essential and that small farms are among the best ways to achieve a sustainable agriculture. The report also questions the sustainability of modern, industrial

---

9. http://www.neh.gov/about/awards/jefferson-lecture/wendell-e-berry-biography. Accessed January 2014.

agriculture. Tilman (2012) claims that agricultural land clearing, climate change, and pollution are "rapidly increasing threats to global biodiversity." Earlier (2002) he reported that "agriculturalists are the principal managers of usable lands and will shape, perhaps irreversibly the surface of earth in the coming decades." Agriculture has long been and remains the dominant, most widespread human interaction with the environment. The projected growth of human population from the present 7.2 to 9 or more billion over the next several decades poses, in Tilman's view (2002), a major challenge to agriculture's practitioners, perhaps the most important they have ever faced. Can all be fed while maintaining terrestrial and aquatic ecosystems on which all depend? Many agree with Tilman's view and strongly suggest that modern industrial agriculture harms biodiversity. "It is true that pesticides and fertilizer tend to reduce the number of species where ever they are used, but intensive agriculture employs less land than extensive farming to produce the same amount of food" (Economist, 2013).

Some of the continuing debate over how to feed a growing population and maintain the ecosystem focuses on two farming systems: *land-sparing*, which separates conservation (not on farm land) and production (on farms) and, it is claimed, produces more food on less land from *land sharing*, which integrates them. Modern agriculture's proponents, generally Malthusian skeptics, advocate land-sparing—large-scale intensive, highly productive farming, which because of higher production per unit of land, can include, when possible, protection of natural habitats from conversion to agriculture. Supporters of land-sharing—small-scale ecologically based agriculture, advocate integrating both objectives (production and protection) on the same land. Some suggest that land-sparing produces more food and is more ecologically beneficial than land-sharing (i.e., small farms). The reports conclude that land-sparing—modern industrial agriculture—is more productive, more beneficial to biodiversity, and maintenance of habitat for other species. Other studies (e.g., Phalan et al., 2011) suggest that land-sharing is the best system. Controversy has arisen because many exclusively focus on production rather than food security, trade-offs, and biodiversity (Fischer et al., 2014). The debate goes on.

Increasing Percentage of Farm Production by Fewer, Large Farms

Given current, sustained trends, and the view of proponents of modern industrial agriculture, it is highly likely that large farms will continue to produce the bulk of US crops and livestock. The 2006 Agricultural Resource Management Survey[10] showed that about 75% of all farms each sold less than $50,000 worth of agricultural products; less than 6% of total agricultural sales. They operated

---

10. http://www.ers.usda.gov/publications/eib-economic-information-bulletin/eib49.aspx#.UjsdQoZJNac. Accessed January 2014.

about 25% of farm acres, had less than 15% of expenses for all US farms, and more than 440,000 (about 1in 5) had less than $1000 in sales in 2006. The increasing dominance of large-scale industrial farms is confirmed by the evidence that fewer than 10% of US farms operated more than 40% of all farm acres, had at least $250,000 in sales in 2006, which equaled more than 75% of US gross agricultural sales. However, 87% of US farms are owned, managed, and operated by individuals or families, who may have incorporated their business. Corporate farms that are owned, managed, and operated by corporate employees account are only 4% of all US farms.

Large farms obtain economies of scale and savings from labor-saving technologies. They achieve the dominant goal of large businesses in industrial economies—maximum, efficient production for short-term economic gain (Kirschenmann, 2014). About 188,000 of the 2.2 million farms in the United States accounted for 63% of sales of agricultural products in 2007.[11] They fulfill agriculture's primary ethical and practical responsibility: continuing and maintaining production of abundant, safe, relatively low-cost food. The results of the evolution in modern American farming are centralization, an absence of open markets, decision making by distant officials, and growing dependence on technology. America's agricultural monopolization is justified by one word—efficiency—delivery of abundant, safe, high-quality produce at the lowest possible cost (Pyle, 2005).

The increasing size of farms and the industrialization of agriculture is "driven by consumer and processor needs and supported by new technology (Urban, 1991)." Uniformity and predictability are essential to efficiency. Farmers feel the pressure to increase the size of their operation, obtain market share, and link with agribusiness to produce and market commodities. Farmers are not the only unique business people in the world, but they are commonly burdened with an obligation not shared by most businesses—pay the price someone else demands for the technology they need and accept the price someone else offers for the commodity they produce.

Farmers have always been interested in protecting and maintaining the productivity of their land—in other words in sustainability. But they operate in a world of short-term economic pressure, which compels increasing production, increasing size, and may necessitate ignoring or at least diminishing the importance of their commitment to stewardship (Hamilton, 1994). It is reasonable to claim that all farmers want to farm so they can farm again tomorrow but the industrial system often makes that difficult, if not impossible.

Small-scale farmers who bring produce to farmer's markets, know that producing high-value food products (e.g., meat, fruit, vegetables) is profitable because it allows the producer to establish the price. Large farms, in a real sense, do not produce food. They produce a saleable commodity for which, as mentioned above, someone else sets the price.

---

11. http://www.epsa.gov/agriculture/ag101/demographics.html. Accessed August 2013, data last updated April 15, 2013.

Increasing Dependence on Technology

The ease of use and low risk of failure of modern technology, especially chemical technology (e.g., pesticides and fertilizers) and the reliability of mechanical technology explains why farmers use them. Uncontrollable events such as drought, severe sudden storms, poor markets, etc. encourage farmers to seek certainty when they can (Zimdahl, 2012, p. 132) and modern technology provides more predictable certainty.

Kroma and Flora (2003) show how the "greening" of advertising in the agricultural media (i.e., farm magazines, radio, television) appropriated current societal values in the imagery that depicts the reliability and environmental acceptability of agricultural products, especially pesticides. Their work demonstrates how pesticide advertising changed from 1940 to 1990 in response to US socio-cultural changes. The industry strategically repositioned itself "to sustain market share and corporate profit by co-opting dominant cultural themes at specific historical moments," and by appearing to adopt expected professional ethical norms (Kroma and Flora). Simultaneously, industry advertising avoided or masked environmental and social challenges to pesticide use. Pest control was accomplished as companies successfully sold farmers new chemical products (Mohler et al., 2001). But farmers are not blind consumers of the technology featured in glossy, well-done advertisements. Pesticides are widely used because they provide predictable results; they work.

AGRICULTURAL EDUCATION

The five characteristics described above have interacted with and influenced how chemistry has transformed agriculture. The modern agricultural system has been created by research in US Colleges of Agriculture in the nation's land-grant universities and complementary, often cooperative, research by agribusiness companies. This unique system has produced some desirable intended and undesirable unintended results. It has maintained food and fiber production while gradually worsening the long-term health of soil and groundwater. Plant and animal genetic diversity have been reduced in a political and economic climate that has reduced crop and livestock choices. The US diet favors animal over plant products that, when combined with poverty, has resulted in society in which 69% of the adults older than age 20 are overweight and 35% are obese and at least 5% are extremely obese.[12] Our capital, energy, and chemically intensive production system requires high production volume at low cost and has driven small- and medium-sized farms out of business (Goldschmidt, 1998). One of the unexpected results is that more developmental research is now done by agricultural industries rather than by scientists in land-grant universities.

---

12. http://www.cdc.gov/nchs/fastats/obesity-overweight.htm. Accessed June, 2014. http://www.win.niddk.nih.gov/STATISTICS/. Accessed July 2014.

THE MORRILL ACT

The greatest, enduring agricultural educational achievement in the United States was passage of the Morrill Act by the US Congress in 1862. The Act of 1862 is well known. The second, the Morrill Land Grant Act of 1890, also known as the Agricultural College Act of 1890, created agricultural colleges and mechanical curricula for former slaves who were unable to gain entrance to white colleges and universities. The 1890 act led to the creation of 17 historically black land-grant colleges in the former Confederate states, which had the unintended consequence of buttressing racial segregation in higher education. States either had to admit freed slaves to existing land-grant colleges and universities or create new postsecondary institutions for qualified students. They often chose the latter option.

During the mid- to late 1800s, there was support among progressive farmers for agricultural education and agricultural colleges. Few of the colleges established before the Morrill Act achieved the expectations of farmers or politicians and their failure gave impetus to those in favor of the Morrill Act. Four states had established agricultural colleges prior to passage of the Morrill acts. Michigan created the Agricultural College of the State of Michigan in 1855. Pennsylvania State University began as the Farmer's High School in 1855. The College of Agricultural Sciences was the first college and awarded the nation's first baccalaureate degrees (13) in agriculture in 1861. The University of Maryland was chartered in 1856 as the Maryland Agricultural College. In 1858, the Iowa legislature established the State Agricultural College and Model Farm. It was the first state to accept the provisions of the Morrill Act by action of the legislature on September 11, 1862.

It is generally and incorrectly assumed that Senator Justin S. Morrill of Vermont originated the plan for creation of land-grant colleges in each state. Wiest (1923, p. 193) cites an article written by Edmund James (1910), the president of the University of Illinois, in which James gives credit for the idea to Prof. Jonathan Baldwin Turner who taught at Illinois College in Jacksonville, IL, from 1833 to 1848. Turner had studied at Yale and was a farmer in Illinois as well as a teacher. According to Wiest (1923, p. 193):

> In 1851 in order to head off a movement to divide a fund amounting
> to about $150,000 and known as the college and seminary fund,
> among the private colleges in Illinois, the farmers by public notice
> at county fairs and in the press were called to meet in convention to
> consider "such measures as might be deemed most expedient to
> further the interests of the agricultural community, and particularly
> to take steps toward the establishment of an agricultural university".

Turner was the leading spirit of the convention and drew up resolutions asking that an industrial university be established in each state of the Union.

Two more conventions were held in 1852 and a fourth in January 1853, in which a petition was drafted to be approved by the Illinois legislature and forwarded to the US Congress:

> ...for the purpose of obtaining a grant of public lands to establish and endow industrial institutions in each and every State in the Union.

In 1854, friends suggested Justin Morrill as a candidate for the US House of Representatives. In December 1855, as he began his first term in the House, Turner's plan had been discussed throughout the country, and Morrill, although not the originator of the plan, became its champion when Turner selected him to propose his idea to Congress. Morrill is given the credit for the act that bears his name, but it was Turner's idea that created the Morrill Act. The idea of using public land to support education had been common practice from the earliest colonial days (Wilson and Fane, 1966, p. 4). The proposal to make it a nation-wide program was new.

Morrill introduced the land-grant bill in 1857, the beginning of his second term in the House. The majority report on the bill was unfavorable. It was passed in the House by a five-vote margin in 1858. In 1859, it was introduced in the Senate, where it was vigorously opposed by southern Senators who regarded it as an attempt by the federal government to tell each state what it had to do with its own land. The bill violated what southern Senators regarded as states' rights. The bill passed the Senate by a margin of three votes, but was vetoed by President James Buchanan (a native of Pennsylvania and the only US president who never married) ostensibly because it was too expensive and unconstitutional. Buchanan, allegedly also faced pressure from Southern senators and representatives who were opposed to the bill. It was reintroduced by Senator Wade of Ohio in 1862 and passed 32 to 7. President Lincoln signed it on July 2, 1862, in the midst of and largely because of the Civil War, which because of the secession of the southern states eliminated the opposition of southern senators. The Republican majority in the Congress passed the Morrill Act and two other significant pieces of legislation that, although not specifically agricultural, had major effects on agriculture. The first was the Homestead Act, which promised 160 acres of free public land, mainly in the west, to anyone who settled on the land for at least 5 years. The second was the Pacific Railroad Act that provided funds for construction of the first trans-continental railroad, which greatly encouraged western agricultural development.

Five Primary Features of the Morrill Act

1. Each state received a grant of 30,000 acres of federal land for each senator and representative, excluding mineral lands. The largest land appropriation was 990,000 acres in Wisconsin to benefit Cornell University in Ithaca, NY. The total land allocation to all states was 17,400,000 acres.
2. In states with no public land, scrip was issued that represented claims to public land that lay elsewhere (e.g., Rhode Island received the benefit of

120,000 acres of land in Kansas). The scrip was to be sold and the proceeds were to be properly invested. The recipient state could sell the scrip or locate, claim, and sell the land authorized. No state could permanently hold land outside its borders to protect states from extending their borders or setting up enclaves in another state.

3. Expenses involved with the management and sale of the land were to be paid by each State and not deducted from the proceeds of sale.

4. The proceeds were to be invested in "stocks of the United States or of the States, or some safe stocks, yielding not less than five per centum upon the par value of the stock." In Morrill's words "The moneys so invested shall constitute a perpetual fund ..." the interest of which was to be appropriated by each state, "which may take the claim and benefit of this act, to the endowment, support, and maintenance of at least one college where the leading object shall be, without excluding, other scientific and classical studies, and including military tactics, to teach such branches of learning, as are related to agriculture, and the mechanic arts, in such manner as the Legislatures of the States may respectively prescribe, in order to promote the liberal and practical education of the industrial classes in the several pursuits and professions in life."

5. The states were responsible for maintaining the funds invested. The act allowed expenditure of up to 10% for purchase of college sites and experimental farms but prohibited use of the funds for construction of buildings or building maintenance. Finally, the act required states that took advantage of the act to create at least one college within 5 years. States in rebellion against the United States were excluded from benefitting from the act while in rebellion.

In some cases, the funds were assigned to private, existing institutions: for example, the Massachusetts Institute of Technology, Amherst College in Massachusetts, and Cornell in New York state. The Massachusetts Institute of Technology and Cornell continue to receive land-grant funds; Amherst does not (Carstensen, 1962). Iowa, the first state to accept provisions of the Morrill Act, was the first state to begin an agricultural extension effort when Perry G. Holden hosted a farmer's institute at Hull, IA, in 1903. Dartmouth, Yale, and Rutgers were also beneficiaries of the Morrill Act, but only the latter created an agricultural college. Brown University was the land-grant college in Rhode Island from 1863 to 1894. Rhode Island had no public lands within its borders to respond to the government's offer. Therefore, the Rhode Island General Assembly of 1863 authorized the governor to accept and receive the funds from the sale of 120,000 acres of land in Kansas under the terms and conditions of the Morrill Act and assigned the funds to Brown University. The state legislature transferred the title and funds to the Rhode Island College of Agriculture and Mechanic Arts in Kingston, RI, in 1892. Later it became the University of Rhode Island (Wilson and Fane, 1966, pp. 3–4). The political machinations were intense. Brown University tried to acquire the agricultural college but its efforts failed when the legislature ended Brown's attempts on May 4, 1892.

Reasons for Initial Failure

For many years, the land-grant agricultural colleges did not fulfill the mission for which they were created. In fact, most failed. The primary goal was education of a modern farmer who would be a better and more efficient manager of farm resources (Campbell et al., 1999, p. 91). No one knew exactly how to do this. The land-grant colleges did not fail for lack of trying but for seven understandable reasons.[13] The reasons were enumerated by Dean Emeritus Eugene Davenport of the University of Illinois College of Agriculture in an address published in the Proceedings of the 26th Annual Convention of the Association of American Agricultural Colleges and Experiment Stations held in Atlanta, GA, in 1912.

It Was a Radical, Absurd Education Experiment

The creation of new agricultural colleges in each state was a radical departure. Although the creators knew what they intended, the purposes, rationale, and methods were not clear to all. Inevitable mistakes and poor communication with the public led to doubt by supporters and opposition of enemies. Many saw the whole enterprise as absurd primarily because of the opponents' surety that mixing education and industry (the mechanic arts) was impossible. It was not unusual to hear the claim that "book learning" was not the proper or even a feasible way to learn how to farm.

It Was Opposed by the Classicists—The Educational Establishment

There was special and clear opposition from those in liberal arts colleges (the classicists) who never saw education as having any applied purpose. Education was based on study of the classics, and prepared men (and a few women) for the learned professions: the church, law, or medicine. To create educational institutions specifically designed to teach practical things was antithetical to the purpose of higher education. Land-grant institutions would lower educational standards and commercialize education, which should not be allowed. This is a problem that many argue exists today. Most land-grant institutions give a prominent place to applied research in the physical and biological sciences and engineering, but neglect the humanities. The reasons for this emphasis are clear. All academic institutions need outside funds to exist. Those funds, although not abundant are more available in science and engineering than in the humanities. Thus, acquiescence to the well-defined needs of funding agencies in the interest of national defense or business development and income from patents is understandable but does make one sanguine about the future of liberal education in land grant institutions.

---

13. Free Press Printing Co. 1913. Burlington, VT. 26:156–166. Also see Wiest (1923), pp. 207–208.

It Was Opposed by Farmers

The opposite reaction came from farmers the land-grant institution was designed to assist. The essence of their objection resonates today. There is little justification for those who learn about farming from books to believe they can teach those who acquired farming knowledge and ability from the experience of farming and from the generations that preceded him or her on the land. Books can't teach anyone how to farm; only farmers can. Farmers often thought of a college education as a way to lure their sons to escape from the farm. The few students who were attracted to the land-grant college were also discouraged because they were required to work on the institution's farm as part of their learning. They had to work and pay to attend the school.

There Was Little to Teach

A significant problem and a frequent reason for failure was that those who taught did not have a body of knowledge to teach. In addition, and equally important, there were no trained teachers. There was little to teach and few to teach. A land-grant institution could hire a scientist (usually a chemist) who knew little, if anything, about agriculture to teach agriculture or hire a good farmer who had no teaching experience. There were few textbooks and little recorded research from which one could create a body of teachable knowledge. Education was classical with emphasis on what we now call the liberal arts. Science was only beginning to obtain a place in the academy. In the 1870s, the Michigan Agricultural College had the largest chemical laboratory and the most instruction in chemistry of any institution west of Harvard College. There was no agricultural science or an agricultural research system that asked questions about and stimulated improvement in the practice of agriculture. In the academy, science had a poor reputation and agricultural science, if available, was regarded as nearly illegitimate. Agriculture in the nineteenth century and before was a handicraft in which farmers knew how to farm their land because they had done it, science was not involved and looked upon with suspicion. That does not mean farmers always farmed well.

There Was a Hidden Agenda

A crucial objection, supported by little empirical evidence, but in retrospect surely true, was that the colleges were very successful at training young men (in the early days, all students were men) to leave the farm and find a job in business, often a farm-related business. The essence of the objection was that the colleges were not truthful. Their hidden objection was that agricultural instruction was simply a façade designed to persuade impressionable young men that a better future lay elsewhere. However, in the early days, most students returned to the farm.

It Was Going to be Expensive and Unnecessary

Finally a familiar problem became apparent—effective agricultural instruction was going to be expensive, more expensive than its supporters had anticipated.

More money for people and material was required and the public was regarded as the logical source. In the nineteenth century, as now, many objected to higher taxes to support an enterprise that they thought was unnecessary.

Nevertheless the Morrill Act succeeded "in promoting the education of the industrial classes in the several pursuits of life." It succeeded "in promoting a sound and prosperous agriculture and rural life as indispensable to …national prosperity and security." It was a magnificent initiative and an educational venture of unprecedented scope. It is not wrong to claim as Wiest (1923, pp. 216–217) does that the land-grant program, "marks the beginning of one of the most comprehensive, far-reaching, and one might almost say, grandiose schemes for the endowment of higher education ever adopted by any civilized nation." Most land-grant institutions began as agricultural and mechanical colleges; only a few retain that designation. Most have grown to become major universities with broad educational and research programs that extend well beyond their agricultural origin. A few (e.g., the Massachusetts Institute of Technology) are now dominantly engineering institutions with no agricultural program.

Land in the nineteenth century was cheap and abundant with perhaps an average cost of $1.15/acre (Large, 2003, p. 257). It was so cheap and abundant that it was treated like dirt and farmed to exhaustion. Farmers knew they could move on to what was thought of as the endless frontier. Farming on much of the land was not husbandry, it was exploitation. It was what Large (2003, p. 139) termed "a reckless, improvident, and half-barbaric skinning and stripping of the land." Over the years, land-grant colleges have done much to change how agriculture is practiced. But Large's criticism endures. Modern agriculture has achieved maximum, efficient production for short-term economic gain, but, in the view of many, that enviable goal has been achieved with a system that exploits the land and is therefore unsustainable.

THE HATCH AND SMITH–LEVER ACTS

The Board of Regents of the University of California at Berkeley created the Experiment Station in 1873. There were no appropriations by the California legislature until 1877. Connecticut, by state legislation and appropriations, created the first agricultural experiment station as a separate institution in 1875. It became the Connecticut Agricultural Experiment Station in 1877 (True and Clark, 1900). Its demanding mission was to "develop, advance, and disseminate scientific knowledge, improve agricultural productivity and environmental quality, protect plants, and enhance human health and well-being through research for the benefit of Connecticut residents and the nation."

Because of the Hatch Act, over the next several years, other states established experiment stations (Campbell et al., 1999, p. 180). Some were funded by the state and some had additional private funding. William H. Hatch, a congressman from Missouri, saw his proposal as a way to make US agriculture more competitive in world markets. The Hatch Act was intended to establish and maintain a permanent, effective US agricultural industry through the provision of federal funds to create an experiment station associated with each land-grant college of agriculture. It was signed into law on March 2, 1887, by President Grover Cleveland. The Hatch Act was favored by "urban elites to provide a conservative response to demands of populist farmers and to keep food prices down in urban centers" (Busch, 1982). Urban elites thought agricultural experiment stations would deflect the demands of discontented farmers. If the experimental work was successful, and no one was sure it would be, it would keep food prices low and weaken demands of urban workers for higher wages (Busch). Busch also points out that wealthy farmers supported the Hatch Act because it passed the cost of agricultural experimentation from them to the state. Research, if it was successful, would benefit the wealthier, early adopters whose yields and profits would increase at the expense of other farmers, whereas the costs of developing innovations were borne by the state (Busch). The cost of inevitable research failures would also not be borne by innovators, but by the state. Much of the discussion about the Hatch Act before its passage focused on a debate that continues today—was the purpose of experiment stations to do original research or to provide practical information to farmers? There was never much concern over whether or not it was legitimate for the federal and state governments to sponsor research that would directly benefit farmers. The benefits of the research were to redound to all farmers and family farms which were looked upon with favor and regarded as entitled to some benefit from public funds.

The Hatch Act led to the eventual proposal and passage of the Smith–Lever act of 1914 that created the Cooperative Extension Service in all states. The act was named after US Senator Hoke Smith of Georgia and Representative Asbury F. Lever of South Carolina. It was first proposed by Lever and adopted by Smith for the Senate and final version. It was signed into law by President Woodrow Wilson. At least 5 years of wrangling preceded its signing because there was little agreement on how extension was to be conducted and what US Department of Agriculture's role was to be. The act's purpose was to establish a system of cooperative extension services, connected to the land-grant universities, to inform people about current developments in agriculture, home economics, and related subjects. The federal appropriation for cooperative extension is now shared between the states based on a formula.

University and agricultural college faculty supported the establishment of Cooperative Extension primarily because the faculty saw extension as a way to expand the scope of their research and teaching endeavors and avoid or diminish the often time-consuming task of dealing with farmer's questions (Busch, 1982). Left in their "ivory towers," professors could apply science to solve agriculture's manifold production problems. "At the time it appeared to everyone that science and organization would make a better world for all" (Busch). What was not noticed was that the entire system of land-grant colleges, experiment stations, and Cooperative Extension would be so successful that farming would become a business, not a way of life, and the system would create bigger, highly productive farms and drive small-scale farmers, widely regarded as inefficient, from the land (Busch).

THE PRODUCTIVITY OF AGRICULTURE

It is undeniable that the productivity of United States and other developed countries' agriculture is a scientific and technological marvel. Hugh Sidey (1998), a contributing editor of TIME magazine, delivered the 1998 Henry A. Wallace lecture to the H. A. Wallace Institute for Alternative Agriculture (established in 1983) in Greenbelt, MD. Sidey quoted Dumas Malone:

> *The greatness of this country was rooted in the fact that a single farmer could produce an abundance of food the likes of which the world had never seen or imagined, and so free the energies of countless others to do other things. So much of recorded history is about the struggle of individuals and families to feed themselves. That changed dramatically in this country.*

Sidey proposed that the story of the productivity and success of American agriculture is "The greatest story never told." Few, if any, other segments of the American scientific-technological enterprise have amassed such an impressive record of predictive, explanatory, and manipulative success over many years. American agriculture has been a productive marvel and is envied by many other societies where hunger rather than abundance dominates.

Yield records for nine US major crops[14] during the twentieth century show that yield increases have varied from two- to sevenfold (Warren, 1998) and the rate of yield increase was not slowing. No yields decreased. Scientific advances that led to these steady yield increases include development of higher yielding cultivars, synthetic fertilizers, improved insect, weed, and disease control, better soil management, and mechanization. Reilly and Fuglie (1998) extrapolated yields for 11 major crops from 1939–1994 to 2020. A linear growth model indicated annual growth between 0.7% and 1.3%/year for the crops. An optimistic, exponential model predicted growth as high 3%/year. Yield growth per acre has continued to increase. The US corn harvest in 2013 was a new record high—almost 14 billion bushels. The previous record high was 13.1 billion bushels in 2009. Unfortunately, consumer prices were not significantly affected.

Avery (1997) points out that without the yield increases that have occurred since 1960, the world would require an additional 10–12 million square miles (roughly the land area of the United States, the European Union countries, and Brazil combined) for agriculture to achieve present levels of food production. Avery claims that modern high-yield agriculture is not one of the world's problems but rather the solution to providing sufficient food for all, sufficient land for wildlife, and protecting the environment.

---

14. Corn, cotton, peanut, potato, rice, sorghum, soybean, tomato, wheat.

REFERENCES

Avery, D., 1997. Saving the planet with pesticides and biotechnology and European farm reform. In: British Crop Prot. Conf.—Weeds, pp. 3–18.

Busch, L., 1982. History, negotiation, and structure in agricultural research. Journal of Contemporary Ethnography 11, 368–384.

Campbell, C.L., Peterson, P.D., Griffith, C.S., 1999. The Formative Years of Plant Pathology in the United States. APS Press, American Phytopathological Society, St. Paul, MN. 427 pp.

Carstensen, V., 1962. A century of the land-grant colleges. Journal of Higher Education 33, 30–37.

Economist, September 14, 2013. Biodiversity. 16 pp.

Ford, H., 1916. Quotes in Chicago Tribune. May 25, 1916.

Fischer, J., Abson, D.J., Butsic, V., Chappell, M.J., Ekroos, J., Hanspach, J., Kuemmerle, T., Smith, H.G., von Wehrden, H., 2014. Land sparing versus land sharing: moving forward. Conservation Letters 7 (3), 149–157.

Gardner, B.L., 2002. American Agriculture 20th Century: how it flourished and what it cost. Harvard University Press, Cambridge, MA, 388 pp.

Goldschmidt, E., 1998. Conclusion: the urbanization of rural America. In: Thu, K.M., Durrenberger, E.P. (Eds.), Pigs, Profits and Rural Communities. State University of New York Press, Albany, NY, pp. 183–198.

Hamilton, N.D., 1994. The emerging conflict between industrialization and sustainable agriculture. Northern Illinois Law Review 14 (3), 613–657.

Head, T. (Ed.), 2006. Conversations with Carl Sagan. University of Mississippi Press, Jackson, MS, p. 167. See Kalosh, A. Bringing science down to earth, P. 99.

Iacocca, L., Novak, W., 1984. Iacocca: An Autobiography. Bantam Books, New York, 352 pp.

James, E.J., 1910. The origin of the Land Grant Act of 1862 (The so-called Morrill act) and some account of its author Jonathan B. Turner: The University Studies IV(1). 7–32.

Kirschenmann, F., 2014. Is optimization the answer? Leopold Letter 26 (3), 5.

Kolbert, E., 2013. Head count. The New Yorker. October 21, p. 96–99.

Kroma, M.A., Flora, C.B., 2003. Greening pesticides: a historical analysis of the social construction of chemical advertisements. Journal of Agriculture and Human Values 20, 21–35.

Large, E.C., 2003. The Advance of the Fungi. American Phytopathological Society. St. Paul. MN. First published 1940. J. Cape, London, U.K, 488 pp.

Lockerby, P., 2011. Henry Ford - Quote: "History Is bunk." http://www.science20.com/chatter_box/henry_ford_quote_history_bunk-79505 (accessed January 2014).

Mohler, C.L., Liebman, M., Staver, C.P., 2001. Weed management: the broader context. pp. 494–518. In: Liebman, M., Mohler, C.L., Staver, C.P. (Eds.), Ecological Management of Agricultural Weeds. Cambridge University Press, Cambridge, UK, 518 pp.

Phalan, B., Onial, M., Balmford, A., Green, R.E., 2011. Reconciling food production and biodiversity conservation: land sharing and land sparing compared. Science 333 (6047), 1289–1291.

Pyle, G., May/June 2005. Stalin's revenge: American agriculture increasingly resembles a Soviet failure. Orion 11.

Reilly, J.M., Fuglie, K.O., 1998. Future yield growth in field crops: what evidence exists. Soil and Tillage 47, 275–290.

Sidey, H., 1998. The Greatest Story Never Told: The Food Miracle in America. H. A. Wallace Annual Lecture. H. A. Wallace Inst. For Alternative Agric., Greenbelt, MD, 20 pp.

Tilman, D., Summer 2012. Biodiversity & environmental sustainability amid human domination of global ecosystems. Dædalus 108–120.

Tilman, D., Cassman, K.G., Matson, P.A., Naylor, R., Polasky, S., 2002. Agricultural sustainability and intensive production practices. Nature 418, 671–677.

True, A.C., Clark, V.A., 1900. The Agricultural Experiment Stations in the United States; with Contributions from the Association of American Agricultural Colleges and Experiment Stations. U.S. Govt. Print. Office, Washington, D.C, 636 pp.

Urban, T.N., 1991. Choices: The magazine of food, farm & resource issues. Research in Agricultural and Applied Economics 6 (4), 4–6.

Warren, G.F., 1998. Spectacular increases in crop yields in the United States in the twentieth century. Weed Technology 12, 752–760.

Wiest, E., 1923. Agricultural organization in the United States. University of Kentucky Press, Lexington, KY, 618 pp.

Wilson, J.W., Fane, D., 1966. Brown University as the land-grant college of Rhode Island, 1863–1894. Brown University, Providence, Rhode Island.

Zimdahl, R.L., 2012. Agriculture's Ethical Horizon, second ed. Academic Press, London, UK, 274 pp.

Chapter 2

The Characteristics of Modern Agriculture Enabled by Chemicals

AGRICULTURAL DEVELOPMENT

In the world's rich, developed countries, agriculture has evolved through three defining eras and is now rapidly moving into the emerging era of genetic modification/biotechnology. Unfortunately, and disappointing to all agricultural practitioners including scientists, small- and large-sized farmers, ranchers, commercial suppliers of resources, and developers of agricultural technology is the fact that many of the world's people remain under- or malnourished. The challenge continues in spite of the equally startling fact that enough food is produced by the world's farmers and ranchers to feed everyone. Agriculture's practitioners produce 17% more calories/person today than 30 years ago, despite a 70% population increase. This is enough to provide everyone in the world with at least 2720 kilocalories (kcal)/person/day,[1] which is above a minimally adequate diet of 2400 kcal/day.[2]

One must ask, why don't all have adequate food? The threefold answer is (1) many people do not have sufficient land to grow or (2) income to purchase food that (3) may or may not be available. The answer is further complicated by the fact that there is no right to food. No one has a firm, mutually agreed upon moral obligation to feed anyone. Sadly, others may control its availability to the poor who, it seems, will always be among us.

---

1. http://www.worldhunger.org/articles/Learn/world%20hunger%20facts%202002.htm#Does_the_world_produce_enough_food_to_feed_everyone. Accessed September 16, 2013.
2. http://www.fao.org/fileadmin/templates/ess/documents/food_security_statistics/metadata/undernourishment_methodology.pdf.

Six Chemicals That Changed Agriculture. http://dx.doi.org/10.1016/B978-0-12-800561-3.00002-X
Copyright © 2015 Elsevier Inc. All rights reserved.

AGRICULTURE'S ERAS

Settled Agriculture

In Kolbert's (2014) view, agriculture was invented, several times, in different parts of the world. The invention, development, or creation of settled agriculture was almost certainly by women, who, among their tasks, collected edible seeds and began to plant them. The first time may have been in Southern Turkey and parts of central Europe 10,000 to 7000 years ago, when wheat, or its genetic ancestors, was planted rather than gathered. People began to abandon hunting and gathering because they learned they could grow some of what they needed where they were; moving was not required. Over several millennia, settled agriculture began with maize in Mexico, Panama, and Colombia, and rice in China's Yangtze Valley. The discovery that grain could be grown was one of the most significant events in human history and, in Diamond's (1987) view, the worst mistake ever made. The advantage was "an efficient way to get more food for less work." Settled agriculture gradually released people from the necessity of producing or finding food. It gave us the time to accomplish many things—indeed, to flourish. Without it, life for most people would have remained "solitary, poor, nasty, brutish and short" (Hobbes, 1651). Without it, we would not have Shakespeare's plays, Mozart's symphonies, been to the Moon, or have achieved all the other marvelous things that we enjoy. But, it, Diamond (1987) claims, permitted population growth with its significant negative consequences that dominate our future and created agriculture's production challenge.

Blood, Sweat, and Tears Era

Other than permitting settlements and activities independent of food production, settled agriculture led to the blood, sweat, and tears era, in which famine and fatigue were common, inadequate food supplies were frequent and agriculture was inefficient, hard work. Most people were farmers and many farms were small and operated at a subsistence level. Life, as described by the British philosopher Hobbes (1651) was, for most people:

> *wherein men live without other security, than their own strength, and*
> *their own invention shall furnish them....In such conditions there is...*
> *no knowledge of the face of the earth; no account of time; no arts; no*
> *letters, no society; and which is worst of all, continual fear and danger*
> *of violent death; and the life of man, solitary, poor, nasty, brutish, and short.*

The world's population has not always grown as rapidly as it has in the past several decades (Figure 2.1). Even though for many centuries there was very little if any population growth; however, Hobbes' dismal view still characterizes the lives of too many people. On October 29, 2014, the world's population

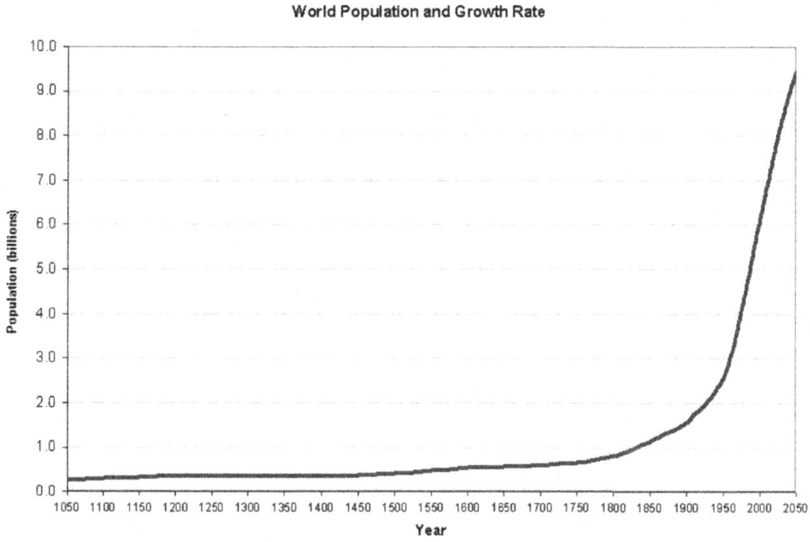

World Population and Growth Rate

FIGURE 2.1 World population growth—1050–2050.

was 7.271 billion, growing at about 1.1%/year (significantly lower than the 1.2–1.3% rate of the previous decade), which presently yields the fact that today the world's farmers and the food system must feed more than 200,000 people than were here yesterday. Growth rate in many African and Middle Eastern countries still exceeds 2%/year.

About half of the world's people live on less than US $2/day. In 2013, the United Nations Food and Agriculture Organization estimated that nearly 870 million of the world's 7.1 billion people, 12%, were hungry. Ninety-eight percent of these people live in underdeveloped (developing is a synonym) countries. Thus, agriculture's practitioners face the continuing production/distribution challenge in the knowledge of the dismal reality that almost one in eight of the people on this earth do not have enough to eat. According to the United Nations Children's Fund (UNICEF, formerly United Nations International Children's Fund), 1.5 million people die each year of hunger, including 16,000 children. They "die quietly in some of the poorest villages on earth, far removed from the scrutiny and the conscience of the world. Being meek and weak in life makes these dying multitudes even more invisible in death.[3]" These are truly dismal statistics about our fellow human beings. It is these people, in so many places, who still live with, indeed endure, the blood, sweat, and tears era of agriculture.

---

3. http://www.globalissues.org/article/26/poverty-facts-and-stats. Accessed September 15, 2013.

The Mechanical Era

The mechanical era of agriculture began with invention of labor-saving machines. In 1793, Eli Whitney invented the first workable cotton gin. Thomas Jefferson made the first moldboard plow in 1794, the cast iron plow was patented in 1797, the first moldboard plow with interchangeable parts (primarily the moldboard) was patented in 1819, and in 1837, John Deere perfected and began to manufacture steel moldboard plows. The new plows didn't just dig and disturb soil, they turned it over, permitted cultivation of harder soil, and permitted agricultural expansion in two ways. They enabled plowing soil and growing crops on land that previously could not be farmed. It was the first machine that allowed one farmer to farm more land.

In 1824, when he was 15, Cyrus McCormick (1809–1884) invented a lightweight cradle for harvesting grain. His father, Robert, saw the potential of a mechanical reaper and worked on a horse-drawn reaper to harvest grain. He was never able to produce a reliable machine. He applied for a patent to claim it as his own invention. By 1831, Cyrus had improved the mechanical reaper. He received a US patent in 1834, founded the McCormick Harvesting Machine Company, which became part of the International Harvester Company in 1902, and began to manufacture the revolutionary machines in 1847. Renewal of his patent in 1848 was denied because Obed Hussey, an American inventor from Maine (1790–1860), had patented a reaper in 1833. Both made and patented several modifications to the reaper. Hussey sold his rights to McCormick in 1858.

The horse-drawn reaper really began mechanized farming. It replaced manual cutting of wheat with scythes, sickles, and cradles with a machine that cut wheat and other small grains more quickly and efficiently and thus permitted one farmer to farm more land. McCormick's reaper cut the grain but did not separate grain from chaff (straw and grain husk). In 1915, International Harvester began to sell the first "combine," a new machine that harvested and threshed wheat in the field.

The plow made agriculture possible in the central plains of North America and in northern Europe. It and the reaper and combine were essential to the creation of modern agriculture. They were part of the industrial revolution, a series of events that freed men from direct dependence on their own and animal energy and, to a degree, freed people from local dependence on the land. The machinery and technology of the industrial revolution made the mechanical revolution of agricultural possible. They permitted long-range, predictable control of yields and thus of food supplies. The industrial and agricultural revolutions were well under way in the United States at the end of the Civil War. They moved side by side. Americans had the advantage of building on the industrial technology and experience of Europeans, where the industrial revolution began. In effect, European mistakes "paid" much of the US development cost. The Morrill Act of 1862 and the Hatch Act of 1887 (see Chapter 1) preceded and became part of America's agricultural revolution, which was aided by expansion of railroads, development of the steel industry, and the opening of Western lands.

TABLE 2.1 Urban[a] and Rural Population of the United States

Year	% Urban	% Rural	% Rural Who Farm
1850	15	85	
1900	40	60	39
1950	60	40	15
1990	75	25	2.6
2000	79	20	1
2010	81	18	1

The Bureau of the Census defines urban as more than 50,000 people, and urban cluster is 2500–50,000. Rural includes all unincorporated areas with population of fewer than 2500.

Each of these events transformed agriculture. In the early 1930s when these things began, half of the American population lived on farms. As the Industrial Revolution proceeded, the need for farm labor declined and people left the farm for the city (see Table 2.1), where they found employment in the factories that made the machines and fertilizers that revolutionized farming. The greater efficiency of mechanized agriculture created demand for industrial labor, which was satisfied as people left the farm for the city where jobs were available. Agriculture's mechanical revolution continued when the internal combustion engine replaced the steam engine in the late twentieth century.[1] Nicholas Otto, a German son of a farmer, invented an effective internal combustion gasoline powered engine in 1876. Henry Ford began Ford Motors in 1903 and began selling tractors with internal combustion engines in 1920.

In the United States in 1900, farmers did chores by hand, plowed with a walking plow, forked hay and milked cows by hand, went to town infrequently on a horse or with a horse-drawn wagon, and had animals that provided power needed for some farm operations. Farmers grew nearly all of their own food, and heated the home with wood, which was cut by axe or saw from the farm. Corn was shelled for animal feed with a hand-cranked machine. All seed and hay were handled by laborious hand labor. Women's work was especially difficult and was never finished. Cooking, canning, washing, ironing, mending, sewing, helping with farm work, and hand-pumping and carrying water all were done by hand almost exactly the way their mothers, grandmothers, and great grandmothers had done the same chores.

Farms had neither running water nor electricity. Electricity did not begin to become available until President Franklin D. Roosevelt created the Rural

---

4. Readers interested in further details of important inventions are referred to Fallows, J. The Atlantic Magazine. pp. 56–68.

Electrification Administration in 1935, as part of the "New Deal." By 1942, about half of all farms had electricity; by 1952, nearly all did. Most farms had tractors (though many still had farm animals), milking machines, and trucks. Many, but not all, had running water. Use of mineral fertilizer increased, and most farmers bought an increasing portion of their food in town. Rural electrification made it possible for farms to have radio, television, telephones, refrigerators, and freezers. It was a new world for farmers and farm wives.

The Chemical Era

The third era of agriculture, the chemical era, boosted production again. The chemical era really began when nitrogen (Chapter 4) and phosphorus (Chapter 5) fertilizers became readily available and increased yields of newly available hybrid corn. Table 2.2 shows the increase in annual fertilizer use from 1900–1909 to 2010. The sharp decline after the decade of the 1980s is due to farmers' recognition of the inefficiency of high rates of use and their pollution potential.

TABLE 2.2 Fertilizer Use in the United States

Decade	Tons of Fertilizer Used/ Year × 1000	Corn Yield bushel/acre
1900–1909	3738	27.3
1910–1919	6117	25.9
1920–1929	6846	26.5
1930–1939	6600	24.2
1940–1949	13,590	31.2
1950–1959	22,341	44.1
1960–1969	32,374	70.4
1970–1979	43,644	86.6
1980–1989	47,411	105.9
1990–1999	21,486	123.3
2000–2009	21,405	132.4
2010	20,843	152.8
2013	21,753	158.8

Source: http://quickstats.nass.usda.gov/results/18B3211B-14D5-330B-B775-FEF0058719C6.
Accessed January 2014. http://usda.mannlib.cornell.edu/usda/current/htrcp/htrcp-04-11-2014.pdf.
Accessed July 2014.

When nitrogen fertilizer was combined with the new hybrid corn varieties, first experimented with by Henry A. Wallace[5] in 1913, yield and fertilizer use went up rapidly (Table 2.2). Wallace's early work led to his founding of the Hi-Bred Corn Company in 1926, which later became Pioneer Hi-Bred. Hybrid corn was popularized by Roswell Garst,[6] an Iowa farmer who recognized its immense potential compared with open-pollinated varieties. He applied his entrepreneurial skills to establish a prosperous seed business and became a leader in educational programs throughout the Corn Belt to convince farmers to adopt hybrid seed and new production methods. He began large-scale production of hybrid seed in 1931.

In 1933, corn sold for 10 cents a bushel and a fraction (1%) of Iowa land was planted with hybrid corn seed. By 1943, 99.5% of Iowa corn was hybrid. In 1933, corn grain yield was 22.8 bushel/acre, about what it was during the Civil War. In 1943, it was 35.4, but by 1981 it had grown to 108.9 bushel/acre (Hyde, 2002). Iowa grain corn yield peaked at 181 bushel/acre in 2004 and was 166 bushel/acre in 2006. Iowa's corn yield has been increasing on an average of 2 bushel/acre/year/and many view 300 bushel/acre as possible.[7] After 1945, when pesticides were developed and became widely available, yields continued to increase.

The chemical era of agriculture developed rapidly after 1945. It began much earlier. The first pesticide was probably elemental sulfur dust used in ancient Sumer about 4500 years ago. In 1000 BC, the Greek poet Homer wrote of pest-averting sulfur. The Rig Veda, one of the four canonical sacred texts (the Vedas, 1700–1100 BCE), of Hinduism is still used in India. It mentions the use of poisonous plants for pest control. Theophrastus, regarded as the father of modern botany (372?–287? BC), reported that trees, especially young trees, could be killed by pouring oil, presumably olive oil, over their roots. The Greek philosopher Democritus (460?–370? BC) suggested that forests could be cleared by sprinkling tree roots with the juice of hemlock in which lupine flowers had been soaked. In the first century BC, the Roman philosopher Cato advocated the use of amurca, the watery residue left after the oil is drained from crushed olives, for weed control (Smith and Secoy, 1975). By the fifteenth century, toxic chemicals such as arsenic, mercury, and lead were being applied to crops to kill pests. In the seventeenth century, nicotine sulfate was extracted from tobacco leaves for use as an insecticide. The nineteenth century saw the introduction of two more natural pesticides, pyrethrum, derived from chrysanthemums, and

---

5. Wallace was the eleventh US Secretary of Agriculture 1933–1940, the thirty-third vice president of the United States, 1941–1945, and Secretary of Commerce 1945–1946.
6. Garst was most well known for hosting Nikita Khrushchev on his farm in Coon Rapids, IA, on September 23, 1959. He sold hybrid seed to the Soviet Union beginning in 1955 and is credited with playing a role in improving US–Soviet communication.
7. http://www.agronext.iastate.edu/corn/production/management/harvest/producing.html. Accessed July 2014.

rotenone, derived from the roots of tropical vegetables. Until the 1950s, arsenic-based insecticides and arsenical herbicides as manufactured by the C.B. Dolge Co. were dominant.

Inorganic Chemicals

Historians tell us of the sack[8] of Carthage by the Romans in 146 BC. The Romans put salt on the soil to prevent crop growth. Later, salt was used as an herbicide in England. Chemicals have been used as herbicides in agriculture for a long time, but their use was sporadic, frequently ineffective, and lacked any scientific base (Smith and Secoy, 1975, 1976).

In 1755, mercurous chloride ($HgCl_2$) was used as a fungicide and seed treatment agent. In 1763, nicotine was used for aphid control. As early as 1803, copper sulfate was used as a foliar spray for diseases. Copper sulfate (blue vitriol) was first used for weed control in 1821. In 1855, sulfuric acid was used in Germany for selective weed control in cereals and onions. The US Army Corps of Engineers used sodium arsenite in 1902 to control water hyacinth in Louisiana.

Desperate potato growers in Ireland tried everything to control the Colorado potato beetle. The story (frequently repeated, but perhaps apocryphal) is that around 1868, one man threw some leftover green paint on his potato plants. It worked. The green pigment in the paint was Paris green, a combination of arsenic and copper, commonly used in paint, fabric, and wallpaper. Farmers, eager for a way to control the potato beetle, diluted it with flour and dusted it on their potatoes or mixed it in water and sprayed it. Paris green (copper acetoarsenite) was a godsend to farmers. In Mann's (2011) view, chemists saw it as something with which they could tinker. If arsenic killed potato beetles, what else would it kill? If Paris green worked for potato beetles, would other chemicals work for other agricultural problems? The answer, of course, was yes. A French scientist found that a solution of copper sulfate and lime killed the causal organism of late blight. Thus, spraying potatoes with Paris green and copper sulfate diminished, but did not solve, the beetle or the blight problem. The use and quest for pesticides had begun.

Bouillie bordelaise (Bordeaux mixture of copper sulfate, lime, and water) was applied to grapevines for the control of downy mildew in the late nineteenth century. It has also been suggested that it was not first applied to control anything. It is a green solution and it is possible that a French viticulturist put it on the vines on the edge of his vineyard to discourage passersby from stealing his grapes. Who wants a grape that has an unnatural green color?

Another passerby, a careful observer, noticed that Bordeaux mixture also turned yellow charlock (mustard) leaves black. That led Bonnet, in France in 1896, to show that a solution of copper sulfate would selectively kill yellow charlock plants growing with cereals. In 1911, Rabaté demonstrated that dilute

---

8. To sack is not a verb that originated with the US National Football League.

sulfuric acid could be used for the same purpose. The discovery that salts of heavy metals might be used for selective weed control led, in the early part of the twentieth century, to weed control research with heavy metal salts by the Frenchmen Bonnett, Martin, and Duclos, and the German Schultz (cited in Crafts and Robbins, 1962, p. 173).

Nearly concurrently, in the United States, Bolley (1908) studied iron sulfate, copper sulfate, copper nitrate, and sodium arsenite for selective control of broadleaved weeds in cereal grains. Henry Luke Purdue University in West Lafayette, IN, awarded a Bachelor of Science degree in 1888 and a Master of Science degree in 1889. In 1890, he was appointed as a botanist and plant pathologist at the North Dakota Agricultural College. He continued to study at North Dakota Agricultural College to become a plant pathologist. Bolley made the first pure cultures of the fungus (*Oospora scabies* Thaxt.), which caused potato scab. The corrosive sublimate treatment for potato scab developed by Bolley became known around the world. Corrosive sublimate is mercuric chloride ($HgCl_2$). It was once used as a treatment for syphilis, but is no longer used for medicinal purposes because of mercury toxicity.

He is acknowledged as the first in the United States to report on selective use of salts of heavy metals as herbicides to eradicate weeds in cereal grains. The action was caustic or burning with little, if any, translocation. It is not an exaggeration to claim that he introduced the idea of broadcast spraying of inorganic herbicides for selective weed control. Although some chemicals were already used to kill weeds and grasses, Bolley believed the Experiment Station should investigate whether chemicals of a sufficient strength could destroy the weeds but not injure cereal grains and beneficial grasses. He began the first US studies with iron sulfate, copper sulfate, copper nitrate, sodium arsenite, and salt for selective control of broadleaved weeds in cereal grains. He believed a tractor sprayer could be driven over the fields to destroy the weeds. Experiments initiated in 1896 were so successful that many states and European countries quickly began spraying cereal crops with inorganic chemicals to increase production. Bolley also corresponded with many manufacturers to help them develop suitable machinery. He knew the available machines (e.g., potato sprinklers) were inadequate and a sprayer with the capacity to create mist under pressure was needed. In North Dakota Agricultural Experiment Station Bulletin 80, published in 1908, Bolley described his view of the future of selective chemical control of weeds.

Each year our experiments have resulted in success of such marked nature that the writer feels safe in asserting that when the farming public have accepted this method of attacking weeds as a regular farm operation that the gain to the country at large will be greater in monetary consideration than that which has been afforded by any single piece of investigation applied to field work in agriculture.

He went on to assert:

If, therefore, this method of attacking weeds by means of chemical
sprays is one-quarter to one-half as successful in general operation
as the writer is willing to vouch for, the money returns to the spring
wheat growing states must far exceed the hopes of the most optimistic.

The advent of 2,4-**dichlorophenoxyacetic** (see Chapter 6) and its many successors has demonstrated the validity of Bolley's prophetic words. He was a pioneer who recognized the potential benefits of selective chemical weed control in the latter part of the nineteenth century, before farmers were ready to adopt it.

He was accurately prophetic about the future of chemical weed control.

Concurrent work in Europe observed the selective herbicidal effects of metallic salt solutions or acids in cereal crops (Zimdahl, 1995). The important early workers with metallic salts were Rabaté in France (1911, 1934), Morettini in Italy (1915), and Korsmo in Norway (1932).

Use of inorganic herbicides developed rapidly in Europe and Great Britain, but not in the United States. In fact, weed control in cereal grains is still more widespread in Europe and England than in the United States. Some of the reasons for slow development in the United States included lack of adequate equipment and frequent failure because the heavy metal salts were dependent on foliar uptake that did not readily occur in the low humidity of the primary grain-growing areas in the United States. The heavy metal salts worked well only with adequate rainfall and high relative humidity. There were other agronomic practices such as increased use of fertilizer, improved tillage (which controlled some weeds), and new varieties that increased crop yield without chemical weed control. In addition, farmers believed that if soils were exhausted and no longer productive, they could move on to the endless frontier. Therefore, initially they were not as interested, as they would be later, in weed control as a yield enhancing technology.

Petroleum oils were introduced in the United States for weed control along irrigation ditches and because, they were selective, in carrots in the early 1900s; although rare, they are still used in some areas. Field bindweed was controlled successfully in France in 1923 with sodium chlorate, which is now used as a soil sterilant in combination with organic herbicides. Arsenic trichloride was introduced as a product called KMG (kill morning glory) in the 1920s. Sulfuric acid was used for weed control in Britain in the 1930s. It was and still is a very good herbicide, but is very corrosive to equipment and harmful to people. In 1940, ammonium sulfamate was introduced for control of woody plants.

The importance of lime and nitrogen and phosphorus fertilizers to the productivity of developed country agriculture cannot be overemphasized. They will be addressed in Chapters 3 (Lime), 4 (Nitrogen), and 5 (Phosphorus).

Organic Chemicals

The chemicals that have done most to transform agriculture are the organic chemicals we know as pesticides.[9] Pesticides are correctly regarded as dangerous poisons, especially if they are used improperly. If they were not poisonous to something, they would not be useful. The prefix, *pest* is derived from the Latin *pestis*. The suffix -*icide* comes from the Latin *caedere*, meaning "to kill." The dictionary definition of a pest is an annoying person or thing, a nuisance, an injurious plant or animal, a pestilence. But exactly what is a pest? There is no widely accepted, precise definition. A pest, therefore, is something that bothers somebody; it is a subjective category. In the agricultural realm, all practitioners know what their particular pests are. Farmers know what insects, weeds, or other pests affect the yield of their crops, the quality of the crop, or the health of an animal.

There are dozens of organic chemicals that are pesticides, more than most people are aware of. This book will not deal with all of them. However, at this point it is appropriate to provide a list of several types of pesticides with brief notation of what they do.

Algaecides are used for killing and/or slowing the growth of algae.

Antimicrobials control germs and microbes such as bacteria and viruses.

Fungicides are used to control fungal problems: molds, mildew, and rust.

Herbicides kill or inhibit the growth of unwanted plants, that is, weeds.

Insecticides are used to control insects.

Miticides control mites that feed on plants and animals.

Molluscicides are designed to control slugs, snails, and other molluscs.

Mothballs are insecticides used to kill fabric pests by fumigation in sealed containers.

Ovicides are used to control eggs of insects and mites.

Rodenticides are used to kill rodents like mice, rats, and gophers.

Other organic chemicals used in agriculture include the following:

Desiccants are used to dry living plant tissues.

Defoliants cause plants to shed their leaves.

Disinfectants control germs and microbes such as bacteria and viruses.

Insect growth regulators disrupt the growth and reproduction of insects.

Pheromones are biologically active chemicals used to attract insects or disrupt their mating behavior. The ratio of chemicals in the mixture is often species-specific.

Plant growth regulators are used to alter the growth of plants. For example, they may induce or delay flowering.

Repellents are designed to repel unwanted pests, often by taste or smell.

Wood preservatives are used to make wood resistant to insects, fungi, and other pests.

---

9. Pesticide is a general term that includes insecticides, herbicides, fungicides, etc.

None of the preceding chemicals is unique to the agricultural world. Many of them are used by gardeners and homeowners to control the creatures they define as pests. Regardless of the user, a condition of use is intentional release into the environment. Pesticides have a well-established, indeed many believe, essential role in agriculture and public health. Agricultural production relies heavily on them. The benefits of their use to assure and improve crop yield, provide favorable economic return, and assure human health and well-being has led to their rapid worldwide adoption.

The first synthetic organic chemical herbicide for selective weed control in cereals was 2-(1-methylpropyl)-4,6-dinitrophenol (dinoseb), introduced in France in 1932 (King, 1966, p. 285). It was used for many years for selective control of some broadleaved weeds and grasses in large seeded crops such as beans. Because of its human toxicity, it was banned by the US Environmental Protection Agency in 1986.

Among the first organic herbicides available in the early 1940s was 4,6-dinitro-*o*-cresol. It was first synthesized in Russia in the mid-1800s and used as a dyestuff, a human sliming agent, and an insecticide (Holly, 1986). Dithiocarbamates were patented as fungicides in 1934.

The effectiveness of monuron (Figure 2.2), a substituted urea, for control of annual and perennial grasses was reported by Bucha and Todd (1951). It was the first of many new selective chemical groups with herbicidal activity. The first triazine herbicide (simazine; Figure 2.3) appeared in 1956 and the first **acetanilide** in 1953 (Zimdahl, 1995, 2010) followed by CDAA, the first alpha-chloroacetamide, in 1956 (Hamm, 1974).

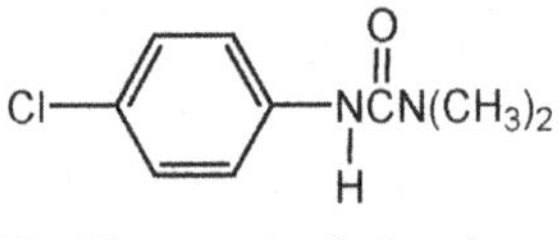

FIGURE 2.2   Monuron. A substituted urea herbicide.

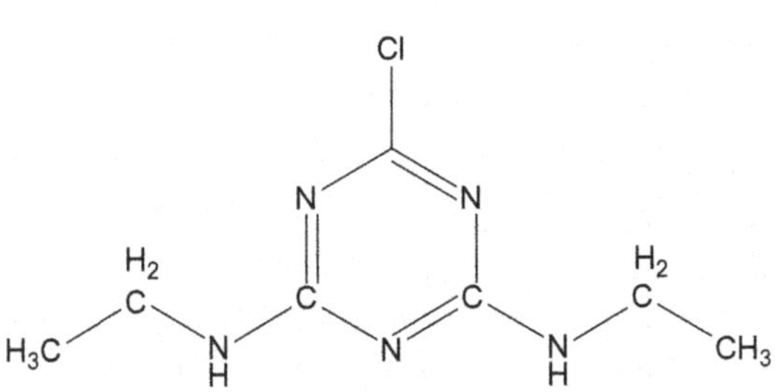

FIGURE 2.3   Chemical structure of simazine—the first triazine herbicide.

Chelates are a class of coordination or complex compounds consisting of a central metal atom attached to a large molecule, called a ligand, in a cyclic or ring structure. They were first used in agriculture in the 1950s to combat plant malnutrition. They are organic chemicals that have the capacity to envelop certain elements (nutrient metal ions, in their agricultural application) and to keep them water soluble under conditions in which they would otherwise become insoluble.

RESULTS AND THE NEW ERA

In 1830, four farmers in the United States supported five nonfarmers. The efficiency of agriculture is reflected in the calculation of the number of people, Americans and non-Americans, fed by one US farmer (Table 2.3). Approximately one in three US crop acres now produce a crop for export.

The efficiency of US agriculture is affirmed by the data in Table 2.4. In 1870, agriculture employed 70–80% of all US workers. In 1900, agriculture employed about a third of all workers, and by 1950 less than a fifth were employed in agriculture. In 2010, of the 2.7 million Americans engaged in production agriculture (farms and ranches), only 1.2 million were owners. Others included hired laborers, workers younger than 20, and younger people.

TABLE 2.3 Number of People Fed by One American Farmer

Year	Number of People Fed	US Population (million)
1910	6	92
1930	10	123
1940	19	132
1950	27	151
1960	61	179
1970	72	203
1980	112	227
1990	122	248
2000	139	281
2010	155	309

Source: http://www.usda.gov/documents/Briefing_on_the_Status_of_Rural_America_Low_Res_Cover_update_map.pdf.

TABLE 2.4 Farm Data 1920–2012

Year	Number of Farmers × 1000	Farmers as % of Labor Force	Acres/Farm
1920	6454	27	148
1930	6295	21	157
1940	6102	18	175
1950	5388	12.2	216
1960	3716	8.3	303
1970	2780	4.6	390
1980	2439	3.4	426
1990	2143	2.6	461
1998	2190	<1	435
2000	2172	<1	434
2012	2170	<1	421

I believe that it is reasonable to claim that the vast majority of the American public (the 99% that does not farm) favors the continuation and support of family farms. Unless they have read Berry's "The Unsettling of America: Culture and Agriculture" (1977) or the Leonard's "The Meat Racket" (2014), it is doubtful that the public has any idea that small family farms are disappearing rapidly from the American landscape. It is true that US Department of Agriculture Secretary Earl Butz's mid-1970s suggestion that farmers "should get big or get out" has been heeded, but I suggest, reluctantly. Small family farms used to be the backbone of the agricultural economy were regarded as a repository and protector of important American values. It is disturbing to many who are directly involved in agriculture and a comment on societal values that the number of American farmers in 2012 was about one-third the number of prisoners in US adult correctional facilities. The disappearance of small family farms is disturbing because it has been caused; it is not a natural phenomenon (Leonard, 2014; Chapter 3). It is also disturbing to know that farmers are growing older. The average farmer was 39 in 1945, 45 in 1974, and 57 in 2012. In 2012, 40% of American farmers were older than 55. One must ask who will be our future farmers as present farmers die. These characteristics of modern farming are not limited to the United States (Economist, 2014). In 1990, about 2.6% of US citizens were farmers (about 2.8 million) (Tables 2.1 and 2.4). Because of advances in technology, developed country agriculture is now in the era of extensive and intensive use of chemical fertilizers and pesticides and is moving rapidly toward the next era of agriculture—the era of biotechnology (see Chapter 9).

Less than 1% of US citizens farm and they produce more than their grandfathers and great-grandfathers ever dreamed possible. Although the number of farmers has declined in nearly every US state, production has increased. Beginning in 2010, the US Department of Agriculture farm census included residential/lifestyle small farms whose operators report a major occupation other than farming, average farm size decreased from its peak of 461 in 1990 to 421 acres in 2012 (Table 2.4).[10] Ninety-seven percent of US farms are family or individual farms; many are very large. Three percent of US farms are corporate farms that capture 28% of sales and government payments (USDA, 2002). US Department of Agriculture data show that just 187,816 of the 2.1 million farmers receive 63% of all farm sales.

It has not always been so. In about 1938, during the Great Depression, if one traveled from Blue Earth to Austin, MN, about 50 miles, one saw many signs of low rent for a farm if the renter would sign a contract to buy the farm after 3 years. The offer from banks and insurance companies that had foreclosed on the farms included loans at 2% interest to buy equipment, seeds, and other farming necessities. Altruistic motives of banks and insurance companies were surely overshadowed by their economic interests. Nevertheless their actions helped create family farms. Similar programs are highly unlikely in the twenty-first century.

These changes are not unique to American agriculture. In 1938, in Great Britain, a million people produced one-third of the food needed for a nation of 48 million. In 1988, only 450,000 British farmers and farm workers produced three-quarters of the food for 58 million people (Malcolm, 1993). Production from each British agricultural worker increased at about twice the rate of increase for the rest of the economy (Malcolm). Less than 3% of the population of Germany works on farms. Farmers account for less than 2% of Europe's working population.

Leonard (2014, p. 10) claims that American agriculture sector is "one of the richest, most productive moneymaking machines in American life." For him, the critical question is not whether there is money in agriculture; it is, who gets it? As agriculture has moved through three eras, one result is the definition of the American farm has changed. The revered small, self-sufficient farms that were owned and run by middle-class families have almost disappeared. The family farm as an independent, self-supporting entity, and a cultural icon is dying. Now few Americans are aware of the values (e.g., help your neighbors, be kind to animals, help those in need, respect the family, respect your elders) that are derived from the agricultural legacy that came from family farms and is the heart of the Jeffersonian agrarian tradition. However, we still admire the values, even

---

10. http://www.nass.usda.gov/Publications/Todays_Reports/reports/fnlo0211.pdf. Accessed January 2014.

though most do not know their origin. The values of rural America family farms no longer serve as an immediate, experiential source from which citizens derive social values or moral sustenance (Zimdahl, 2012). The slow, inexorable, and largely unnoticed change began in the 1950s. Family farms were no longer the economic foundation of families and communities; they were becoming what Leonard (2014) calls rural factories that provide as much food as possible to the large, complicated American food system. Berry (1977) clearly points out that the abundant production of American agriculture is illusory. It is illusory because it does not safeguard its producers, it destroys them. It is ironic that the abundant food we all expect to enjoy is gradually destroying the sustainable agricultural productive base and replacing it with a high-tech capital, energy, and chemical production system that is ever more centralized and controlled by the nonfarming sector of our economy.

Every era "has a theory of rising and falling, of growth and decay, of bloom and wilt: a theory of nature" (Lepore, 2014). This brief chapter recounts a bit of what has happened in the chemical realm to create the present US agricultural system. It is a factual, descriptive, not a theoretical or prescriptive account. It may border on, but is not intended to be, historicism. It does not assume that all "events in historical time can be explained by prior events" (Lepore). If progress is defined as a reduction of the work force, invention and adoption of improved chemical and mechanical technology, and greatly increased production, there is no question that US agriculture has progressed through its three eras and production will increase as the biotech era emerges. Enormous productive improvement occurred in the nineteenth century and continued in the twentieth. The latter part of the twentieth century saw continued growth and more innovative technology, which continues in the present century. However, when one considers, as one ought to, the number of hungry people, the loss of small farms and rural communities, the neglect of externalities, and significant questions about the sustainability of the highly productive US agricultural system, one must question if the system is as successful as its proponents believe it is.

REFERENCES

Berry, W., 1977. The Unsettling of America: Culture and Agriculture. Sierra Club Books, San Francisco, CA. 228 pp.

Bolley, H.L., 1908. Weeds and methods of eradication and weed control by means of chemical sprays. N. Dak. Agric. Coll. Exp. Stn. Bul. 80, 511–574.

Bucha, H.C., Todd, C.W., 1951. 3-(p-chlorophenyl)-1,1-dimethylurea–A new herbicide. Science 114, 493–494.

Crafts, A.S., Robbins, W.W., 1962. Weed Control: A Textbook and Manual, third ed. McGraw-Hill, New York. 660 pp.

Diamond, J., May 1987. The worst mistake in the history of the human race. Discover pp. 64–66.

Economist, May 3, 2014. Bring Back the Landlords. pp. 39–40.

Hamm, P.C., 1974. Discovery, development, and current status of the chloroacetamide herbicides. Weed Sci. 22, 541–545.

Hobbes, T., 1651. Leviathan. Oxford: Clarendon Press Part 1, Chapter 13.

Holly, K., 1986. Herbicides – past, present and future. SPAN 29 (s), 89–91.

Hyde, J., 2002. Four Iowans who fed the world. http://hoover.archives.gov/programs/4Iowans/Hyde-Culver.html.

King, L.J., 1966. Weeds of the World: Biology and Control. Interscience Pub., Inc., New York. 526 pp.

Kolbert, E., July 28, 2014. Stone Soup: How the Paleolithic Life Style Got Trendy. The New Yorker. pp. 26–29.

Leonard, C., 2014. The Meat Racket – the Secret Takeover of America's Food Business. Simon & Schuster, New York. 370 pp.

Lepore, J., June 23, 2014. The Disruption Machine. The New Yorker. pp. 30–36.

Malcolm, J., 1993. The farmer's need for agrochemicals. In: Gareth Jones, J. (Ed.), Agriculture and the Environment. E. Horwood Pub., London, UK, pp. 3–9.

Mann, C., 2011. The eyes have it. Smithsonian 42 (7), 86–106.

Smith, A.E., Secoy, D.M., 1975. Forerunners of pesticides in classical Greece and Rome. Journal of Agricultural and Food Chemistry 23, 1050–1055.

Smith, A.E., Secoy, D.M., 1976. Early chemical control of weeds in Europe. Weed Science 24, 594–597.

USDA, 2002. National Agricultural Statistics Service. NASS Quick Facts from the 2002 Census of Agriculture. http://www.nass.usda.gov/census/census02/quickfacts/organization.htm.

Zimdahl, R.L., 2010. A History of Weed Science in the United States. Elsevier Insights, London, UK. 207 pp.

Zimdahl, R.L., 2012. Agriculture's Ethical Horizon. Elsevier, London, UK. 274 pp.

Zimdahl, R.L., 1995. Introduction. In: Smith, A.E. (Ed.), Handbook of Weed Management Systems. M. Dekker, Inc., New York. pp. 1–18.

Chapter 3

Lime: A Soil Amendment

Chapter Outline

AGRICULTURAL LIME

Limestone is a sedimentary rock composed of different crystal forms of calcium carbonate ($CaCO_3$). It is about 10% of all sedimentary rock. Most also contains skeletal fragments of marine organisms. Historic uses of limestone included mortar and pulverized limestone used to neutralize acidic soils. Burnt lime (CaO, quicklime) is made by the thermal decomposition of naturally occurring things that contain $CaCO_3$ (e.g., limestone, seashells), When calcium carbonate is heating above 825°C (1517°F), calcination or lime-burning liberates carbon dioxide and producing quicklime.[1]

$$CaCO_3(s) \rightarrow CaO(s) + CO_2(g)$$

The primary agricultural use was and is to raise soil pH. There is no reliable record of when lime was first used as a soil amendment. Lime mortar dated 15,000 to 7000 years BCE has been recovered from terrazzo floors in Turkey. Limestone was used to build portions of the Great Wall of China and the Great Pyramid of Giza, Egypt. Lime has many other uses:

calcium supplement for animal feeds;

construction aggregate as a roadway base;

manufacturing of some kinds of glass;

additive to paper, plastics, paint, tiles, and other materials as both white pigment and a cheap filler;

toothpaste;

food supplement as a source of calcium; and an

additive to some pharmaceuticals and cosmetics.

---

1. http://en.wikipedia.org/wiki/Calcium_oxide. Accessed October 31, 2014.

Six Chemicals That Changed Agriculture. http://dx.doi.org/10.1016/B978-0-12-800561-3.00003-1

The world's soils vary in color, texture, structure, and chemical, physical, and biological composition. It is reasonable to claim that soil is one of the most important things on the earth. It is unquestionably essential to agriculture. Soil is the medium in which food is grown. It is not, as many think, just dirt. Soils are not uniform, although they may appear to be, especially at the local level, but in reality they can be very different within a few feet. The variety of soils is the result of five soil-forming factors.[2] Each spans a continuum, which results in their being thousands of different soils in the world.

1. **Climate.** The amount, intensity, and timing of precipitation influence soil formation. Seasonal and daily changes in temperature affect moisture, weathering, and leaching. Wind (erosion) redistributes sand and other particles. Seasonal and daily changes in temperature determine rainfall's role, its effects, the rate of biological activity and chemical reactions, and the resulting vegetation.

2. **Biology.** Plants, animals, microorganisms, and humans, independently and collectively, affect soil formation. Plant roots open channels in soil, taproots penetrate deeply fibrous roots near the surface easily decompose and add organic matter. Animals and microorganisms mix soils. Microorganisms affect chemical exchanges between roots and soil. Humans mix soil, often extensively, and grow the plants they want.

3. **Landscape position, topography.** Slope and directional orientation affect soil moisture and temperature. Steep slopes facing the sun are warmer. Slopes may lose topsoil as they form and be thinner than nearly level soils that receive deposits from areas above.

4. **Parent material.** Most soil has been created from materials that have moved in from miles or only a few feet away. Loess, an aeolian (windblown) sediment, is common in the good soils of the midwestern United States and some parts of China. It is formed by the gradual accumulation of wind-blown silt (20–50 μm particles) and has 20% or less clay. Loess soils are typically near neutral pH.

5. **Time.** Soil formation is continuous. Over time, soils exhibit features that reflect the other forming factors.

The primary, if not the only, reason agricultural lime ($CaCO_3$) is added to soil is to raise soil pH toward neutrality (7.0), thereby reducing soil acidity, increasing nutrient availability, and permitting successful growth of many crops that are pH sensitive. pH is an abbreviation for potential hydrogen. It is the negative logarithm (base 10) of the reciprocal of the hydrogen ion concentration in gram atoms/liter of water. It indicates hydrogen ion activity. It defines acidity or basicity (pH aq) of a solution on a scale from 0 to 14. Seven is neutral; below is acidic and above is basic. That means that for each unit pH increases or

---

2. http://staffweb.wilkes.edu/brian.oram/soilformingfactors.html. Accessed July 2014.

decreases, basicity or acidity changes by 10 times. A pH of 5 is 10 times more acidic than pH 6 and 100 times more acidic than pH 7.

Because pH controls many soil chemical processes and the chemical form of nutrients, it may enable or inhibit their uptake. Acid soils (below pH 5.5) have greatly reduced microbial activity, but release many nutrients, notably iron, which is much less available above pH 7.5. Soils with pH lower than 4.6 are too acidic for most plants. Many soils are naturally calcareous[3] (pH above 7). In some cases, sulfur can be added to make them more acidic through formation of sulfuric acid (H_2SO_4), hydrogen sulfite (HSO_3), and hydrogen sulfide (H_2S). At a pH above 7, carbonates and oxides are formed and can react with many metallic nutrient elements (e.g., iron, copper, molybdenum), which renders them insoluble and therefore unavailable to plants. Most crops do not grow well in acidic soil or soil with a pH above 8. Raising the pH of acidic soil improves plant growth, may improve water penetration, and reduce aluminum toxicity. Lime is a source of calcium and magnesium for plants. Because it is high in calcium, it can also be beneficial to bone growth of foraging animals.

Soils in high rainfall areas become acidic through leaching. Crop growth and livestock grazing remove essential nutrients over time and soil may gradually become acidic. Chemical fertilizers required to achieve maximum yield are major contributors to soil acidity. Therefore, liming acidic soil is essential to achieve maximum yield of food crops grown in acidic soils.

In areas of extreme rainfall and high temperature,[4] clays and humus may be leached away, which further reduces soil's buffering capacity (i.e., resistance to changes in pH). In low-rainfall areas, unleached calcium may raise pH to 8.5 and if exchangeable sodium levels are high, soil pH may reach 10. Above pH 9, most food crops will not grow or their growth and yield will be severely reduced. High pH also results in low micronutrient mobility and availability.

The desirable (optimum) pH range for most food crops is 5–7. Every crop has an optimum pH range within which production potential peaks. No important food crops have an optimum pH less than 5. Rye, oats, and lupins are acid-tolerant. The optimum for corn and soybeans is 5–7.5. The most important food crops (beans, rice, and wheat) have an optimum pH between 5 and 7. They are acid sensitive. The 10 most important food crops—the plants that feed the world—all grow best between pH 5.5 and 6.5 (see Table 3.1). Potatoes, an exception, grow when soil pH is 4.8–5.5, although they grow well above pH 5.5. Many plants, but not the important food crops, have adapted to thrive at pH values outside the optimum range, but do best within the optimum.

---

3. http://soils.cals.uidaho.edu/soilorders/maps.htm. Accessed July 2014.

4. Determination of low, high, and extreme rainfall is quite subjective. Global average rainfall over land is 28 inches/year (71 cm). Rainforests typically receive 69–79 inches/year (175–200 cm). Temperature is similar. Hot for a North Dakotan may be just right for an Arizonan.

TABLE 3.1 Optimum pH Range for Some Food Plants[a]

pH Range		
5.0–5.5	**5.8–6.5**	**6.5–7.0**
Blueberries	Bean X	Alfalfa
White potato X	Cassava X	Barley X
Sweet potato X	Corn/maize X	Cherry
Grasses X	Grapes	
Millet X	Soybean X	
Oats	Sugarbeet	
Rice X	Wheat X	
Rye		
Some clovers		
Sorghum X		
Tomato		
Watermelon		

[a]*12 of the world's most important food crops are indicated with X.*

About 30% as much phosphorus is available when pH is below 6 versus above 6.5. Nitrogen and potassium become less available below 6. Availability of both decreases by about 30% at pH 5.5 and 70% at 5. Below pH 5, soil nitrogen, phosphorus, and potassium concentrations may be adequate to support plant growth, but, because of formation of insoluble minerals, they are unavailable and of little use to plants. In contrast, iron, copper, manganese, and zinc are most available when pH is acidic. It seems contradictory, but when pH is around 4, other nutrient levels may be high because lack of plant uptake leads to their accumulation in soil.

The most effective way to raise pH is to apply good-quality, finely ground agricultural lime. Lime's calcium and/or magnesium carbonate content and the fineness of grinding determine quality. Finely ground lime will raise soil pH more rapidly. Incorporating it in soil is more efficient than surface application. Change in pH is more rapid when soil is moist from irrigation or rainfall; water is required for the chemical reaction. Action (pH change) may take a year or more in dry soil. However, with good soil moisture, a response may be observed within weeks if pH is extremely low. When pH is low, it is important to apply lime immediately after the growing season to allow sufficient reaction time before the next crop is planted.

TABLE 3.2 Composition and Calcium Carbonate Equivalent of Agricultural Limes

Calcium		Carbonate
Material	Composition	Equivalent (calcium carbonate equivalent)
Calcitic limestone	$CaCO_3$	85–100
Dolomitic limestone	$CaCO_3$; $MgCO_3$	95–108
Marl	$CaCO_3$	50–90
Hydrated lime	$Ca(OH)_2$	120–135

http://www.plantstress.com/articles/toxicity_m/soilph%20amend.pdf. Accessed February 2014.

SOURCES OF LIME

Limestone, a naturally occurring sedimentary rock is composed of high levels of calcium, or magnesium carbonate, or dolomite,[5] and other minerals. After removal from quarries and mines, the stone is crushed, which may be followed by secondary or tertiary crushing and screening into particles ranging in size from several inches to dust. Water and particle size are important:

Particles that pass a 100-mesh sieve react 100% in 6 months or less.
Particles that pass a 60-mesh or finer sieve react 100% within 1 year.
Particles that pass a 20-mesh, but not a 60-mesh sieve, react about 50% in the first year.
Particles not passing a 20-mesh sieve have little liming value.

Because soil acids are relatively weak and soil colloidal surfaces are more accessible to fine particles, lime's efficacy, its neutralizing value, is dependent on its purity, particle size (fineness of grinding), and its chemical composition (Table 3.2). Table 3.2 uses the calcium carbonate equivalent as a measure of efficacy. When the chemistry of a particular product, its calcium carbonate equivalent is combined with particle size the effective calcium carbonate equivalent is determined. It compares the percentage comparison of a particular lime with pure calcium carbonate with all particles smaller than 60 mesh. Commercial agricultural limes will have effective calcium carbonate equivalents between 45% and 110%.

---

5. Dolomite, first collected in the Dolomite Alps of northern Italy, was first described by the French geologist Déodat G. de Dolomieu (1750–1801). Nicolas de Saussure named it after de Dolomieu in 1792. http://en.wikipedia.org/wiki/Dolomite. In some US states, limestone must contain at least 6% Mg to be classified as dolomite. Accessed February 2014.

Lime is sold as it comes from a mine or further crushed/ground to make hydrated lime. Ground limestone is mostly calcium carbonate and generally less than 1–6% magnesium. When calcitic ($CaCO_3$) or dolomitic lime [$Ca\,Mg(CO_3)_2$] is added to soil, they hydrolyze (decompose in water) to a strong base: calcium hydroxide—$Ca\,(OH)_2$—and carbonic acid—H_2CO_3—a weak acid.

$$CaCO_3 + 2H_2O \Rightarrow Ca(OH)_2 + H_2CO_3$$

Calcium hydroxide rapidly ionizes to CA^{++} and OH^- ions. The calcium ions replace absorbed H ions on the soil colloid and thereby neutralize soil acidity. Carbonic acid slowly and partially ionizes to H^+ and CO_2. The net effect is release of more calcium than hydrogen ions which neutralizes soil acidity.[6]

Hydrated lime—$Ca\,(OH)_2$—is calcium hydroxide, sometimes called slaked or builder's lime. It is powdery, quick-acting, and unpleasant to handle. The neutralizing value ranges from 120 to 135 compared with pure calcium carbonate (85–100). Fifteen hundred pounds of hydrated lime with a neutralizing value of 135 is equivalent to 2000 pounds of agricultural lime (neutralizing value, 100). In spite of its efficacy, it is not in common use because it is so unpleasant to handle.

Marls, calcium carbonate mixed with clay and sand, are found mostly in the Coastal Plains of the Eastern United States. Their neutralizing value (70–90) is reduced by clay impurities. Marls are often plastic, lumpy, usually low in magnesium, must be dried and ground before application, and have a soil reaction similar to calcitic lime.

Calcitic or dolomitic limestone has a calcium carbonate equivalent of 100. If another material has an equivalent of 50, it has half of the soil acid neutralizing value and twice as much should be applied. If a material has an equivalent of 150, then only 67% is needed. Applying calcium carbonate ($CaCO_3$) increases the pH of soil and exchangeable calcium level while neutralizing hydrogen ions. It may slightly improve water penetration when soil pH is below 6, but as pH rises above 7, it has little effect on water penetration. Application of gypsum ($CaSO_4 \cdot 2H_2O$) does not neutralize acidic soil or increase pH. It can be a source of calcium and sulfate sulfur. It is never recommended as a substitute for agricultural lime, although it can be used to reclaim sodic (high sodium) soil by replacing sodium ions with calcium. It may increase water penetration.

Calcium oxide is widely used, but not as agricultural lime. It is a white, caustic, alkaline crystalline substance that is solid at room temperature. Intense illumination—limelight—is created when calcium oxide is heated. Limelight (also known as Drummond light (Thomas Drummond, 1797–1840) or calcium light) was used in theaters around the world in the 1860s and 1870s to highlight solo performers, as modern spotlights do. Although limelights are no longer

6. The Efficient Fertilizer use guide. 2014. Soil pH. The Mosaic Company. http://www.cropnutrition. com/efficient-fertilizer-use-guide. Personal communication. February 2014. C.S. Snyder. Nitrogen Program Director, Int. Plant Nutrition Inst. Conway, AR.

used, the term "in the limelight" is still common. More prosaically, calcium oxide is also used in cement, as an indicator of water in petroleum, in paper production, and in plaster. It causes severe irritation when inhaled or contacts moist skin or eyes. During the reign of Henry III, the English Navy destroyed an invading French fleet by blinding the enemy with quicklime.

WHERE LIME IS NEEDED

Approximately 30% of the world's total land area consists of acid soils. More importantly, as much as 50% of the world's arable land may be acidic (von Uexküll and Mutert, 1995) (Figures 3.1 and 3.2). Therefore, without lime, it is reasonable to conclude that the plants that feed the world (X in Table 3.1) could not be grown successfully on 50% of the world's arable land. Aluminum (Al) in these soils will be solubilized to toxic ionic forms, especially at soil pH below 5 (Shao Jian Zheng, 2010). Ionic forms of Al inhibit root elongation by destroying the cell structure of the root apex, affect water and nutrient uptake, and inhibit plant growth and development, often to the point of crop failure. Aluminum is the major cause of plant toxicity in acid soils with a high mineral content (Samac and Tesfaye, 2003). Phosphorus (P) is readily sorbed (fixed) by clay minerals, including various iron oxides and kaolinite clay, that are abundant in acidic soils, which renders it unavailable for root uptake. Thus, Al toxicity and P deficiency are the main constraints to crop production in acid soils.

To produce an optimum crop yield on acid soils, application of lime (primarily calcium carbonate) is recommended to increase soil pH, decrease or eliminate aluminum toxicity, and increase phosphorus availability. Only 11% of the Earth's soils have no inherent limitations for growing agricultural crops. About 50% of

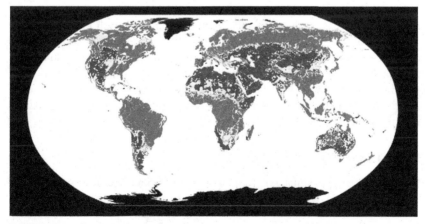

FIGURE 3.1 Global variation in soil pH. Red (light gray in print versions), acidic; yellow (white in print versions), neutral; blue (dark gray in print versions), alkaline; black, no data. http://en.wikipedia.org/wiki/Soil_pH. *Accessed February 2014.*

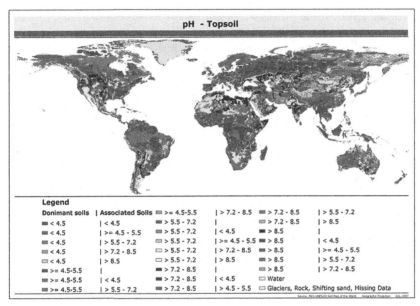

pH - Topsoil

Legend

Donimant soils	Associated Soils	>= 4.5-5.5	> 7.2 - 8.5	> 7.2 - 8.5	> 5.5 - 7.2
< 4.5	< 4.5	> 5.5 - 7.2		> 7.2 - 8.5	> 8.5
< 4.5	>= 4.5 - 5.5	> 5.5 - 7.2	< 4.5	> 8.5	
< 4.5	> 5.5 - 7.2	> 5.5 - 7.2	>= 4.5 - 5.5	> 8.5	< 4.5
< 4.5	> 7.2 - 8.5	> 5.5 - 7.2	> 7.2 - 8.5	> 8.5	>= 4.5 - 5.5
< 4.5	> 8.5	> 5.5 - 7.2	> 8.5	> 8.5	> 5.5 - 7.2
>= 4.5-5.5		> 7.2 - 8.5		> 8.5	> 7.2 - 8.5
>= 4.5-5.5	< 4.5	> 7.2 - 8.5	< 4.5	Water	
>= 4.5-5.5	> 5.5 - 7.2	> 7.2 - 8.5	> 4.5 - 5.5	Glaciers, Rock, Shifting sand, Missing Data	

Source: FAO-UNESCO Soil Map of the World   Geographic Projection   AGL-2007

FIGURE 3.2   The naturally acid and basic world soils. http://www.fao.org/soils-portal/soil-management/management-of-some-problem-soils/acid-soils/en/. *Accessed December 2014.*

the Earth's arable soils are naturally acidic (Northern and Eastern North America, Northern South America, Northern Europe, the former Soviet Union, Central Africa, China, South Pacific, Central Australia, and New Zealand) (Figure 3.1).

Rice, a daily dietary staple for about half of the world's population, is grown in areas of the world with dominantly acidic soils (red in Figure 3.1: Central and West Africa, Brazil, Southern Asia, Pacific island nations (e.g., Philippines, Indonesia, Sri Lanka)). There are large areas of high pH rice soils in important rice growing areas, such as the Indo-Gangetic Plain. A result of soil submergence is that soil pH increases toward neutrality and the pH of soils above 7 decreases to near neutrality. As a result, pH is typically not a serious constraint for flooded rice soils, but it may be for nonflooded (upland rice) soils. The world's potatoes are grown successfully in areas with dominantly acidic soils. Nearly all crops grown in the red areas of Figure 3.1 need lime to produce maximum yield of high-quality food.

EFFECTS OF AGRICULTURAL LIME[7]

Because it decreases soil acidity, application of lime increases fertilizer use efficiency (Table 3.3). It increases availability of nitrogen, phosphorus, potassium,

---

7. Much of this section is an edited version of http://nzic.org.nz/ChemProcesses/soils/2C.pdf. Accessed February 2014.

TABLE 3.3 Percent Fertilizer Use Efficiency at Several pH Levels

Soil Acidity	pH	Nitrogen	Phosphorus	Potassium	% Fertilizer Wasted
Extremely acid	4.5	30	23	33	71.3
Very strong acid	5.0	53	34	52	53.7
Strongly acid	5.5	77	48	77	32.7
Medium acid	6.0	89	52	100	19.7
Neutral	7.0+	100	100	100	None

http://www.plantstress.com/articles/toxicity_m/soilph%20amend.pdf. Soil can have a natural pH below 4.5 (ultra acid) and above neutrality 7. Accessed February 2014.

calcium, sulfur, and magnesium. Applying phosphorus fertilizer will increase available P. However, soil has a huge buffering capacity that diminishes the effects (the efficacy) of all kinds of amendments. Soil acidification is accelerated by acid rain and excess application of ammonia-based, inorganic nitrogen fertilizers. Without pH adjustment (liming), other factors will control crop growth and yield. The beneficial effects of lime application and phosphorus fertilization are usually not permanent; annual application is required. Because application is generally restricted to the soil surface, subsoil properties are hardly modified unless lime is incorporated, thus limiting the effects for crops with deep root systems. Despite all these constraints, abundant vegetation grows on acid soils because many plants have evolved to deal with aluminum toxicity and phosphorus deficiency. Unfortunately, few members of these abundant vegetative communities are food plants.

It was assumed that toxicity of hydrogen ions (hydronium ion $[H_3O^+]$) or calcium and/or magnesium deficiency explained why many plants grew poorly in acidic soil. Research has shown that soil acidity has to be below 3 before the hydronium ion concentration itself is toxic to most crop plants. Calcium and magnesium deficiencies are rare until pH is below 4.0–4.5. Thus, other pH-related effects must explain why plants grow poorly in acidic soils.

Below pH 5.0, aluminum and manganese reach toxic concentrations. Lime reduces the concentration of Al^{3+} and Mn^{2+} in the soil solution, which thereby is favorable for plant growth. Unlike most trace elements, molybdenum availability increases as pH rises, but is below optimum for most plants at pH 5.0; at this pH, molybdenum deficiency may be the most important factor limiting plant growth. Molybdenum deficiency may be corrected for all but some legumes (e.g., alfalfa, peas) by adding lime to raise pH to between 5.5 and 6.5.

US Department of Agriculture scientists in Minnesota (Samac and Tesfaye, 2003; Tesfaye et al., 2001) produced transgenic alfalfa that made it more tolerant of acid soil and less susceptible to aluminum toxicity.

Alfalfa research did not continue; aluminum tolerance was studied in an annual relative of alfalfa (*Medicago truncatula*) (Chandran et al., 2008).

Fertilizers provide, in varying proportions, the six essential macronutrients: nitrogen (N), phosphorus (P), potassium (K), calcium (Ca), magnesium (Mg), and sulfur (S). Larger amounts of macronutrients are required and are present in plant tissue in quantities from 0.15% to 6.0% of dry matter. Eight essential micronutrients: boron (B), chlorine (Cl), copper (Cu), iron (Fe), manganese (Mn), molybdenum (Mo), zinc (Zn), and nickel (Ni) can be supplied by fertilizer. Smaller amounts of micronutrients are required and are present in plant tissue in parts per million, ranging from 0.15 to 400≤0.04%. Four of the macronutrients (N as NH_4^+, K, Ca, and Mg) and four of the micronutrients (Mn, Cu, Fe, and Zn) are absorbed by plants as cations; therefore, liming of acidic soils is required to make the essential cationic plant nutrients available in appropriate quantities.

Exchangeable bases (Ca^{2+}, NH_4^+, Mg^{2+}, K^+, and Na^+ cations) balance most of the negative charges present on organic matter and clay mineral surfaces. They are in dynamic equilibrium with cations present in the soil solution. When more rain falls than is removed by evaporation or plant transpiration, it can flow through soil, leach water-soluble cations and replace them with acidic $H_3O^+_{(aq)}$ and $AL_3^+_{(aq)}$ ions.

As soil pH rises, levels of available aluminum, manganese and, to a lesser extent, molybdenum are reduced. Liming stimulates soil biological activity, increases cycling of the primary plant macronutrients and micronutrients and thereby increases their availability. By stimulating earthworm activity, liming can increase rainfall infiltration, thus reducing the potential for runoff and erosion while increasing plant-available soil moisture. However, adding lime to increase soil pH above 6.5 may do more harm than good by reducing phosphorus availability and causing deficiencies of micronutrients such as manganese and zinc. Excess lime may also increase the loss of sulfate if excess calcium is leached. Liming to increase soil pH above 6.5 should only be undertaken for crops known to grow best at these pH values (e.g., alfalfa).

Initial decomposition/mineralization of plant litter produces carbon dioxide and converts organic forms of nitrogen, phosphorus, and sulfur to simple inorganic forms that can be used by plants. All of these processes tend to increase soil acidity. Excess carbon dioxide forces the carbonic acid dissociation equilibria to the right, releasing hydronium ions:

$$CO_2(g) + 2H_2O(l) \rightleftharpoons H_3O^+_{(aq)} + HCO_3^-_{(aq)}$$

$$HCO_{3(aq)} + H_2O(l) \rightleftharpoons H_3O^+_{(aq)} + CO_3^{-2}_{(aq)}$$

Organic acids decrease soil pH. Release of hydrogen ions in some mineralization reactions involving nitrogen, phosphorus, and sulfur decreases soil pH. Proteins are degraded by heterotrophic microorganisms $NH_4^+{}_{(aq)}$ or $NH_{3(aq)}$

Nitrosomonas degrades ammonia $2NH_4^+{}_{(aq)} + 3O_2(g) + 2H_2O \rightarrow NO_2^-{}_{(aq)} + 4H_3O^+{}_{(aq)}$ and

Nitrobacter degrades nitrate $2NO_{2(aq)} + O_2(g) \rightarrow 2NO_3^-{}_{(aq)}$

Alkaline soils (pH > 7.0) usually occur in dry regions where $Ca^{2+}{}_{(aq)}$ ions and carbonate anions do not leach away. In very arid areas, even the highly mobile sodium (Na^+ aq) and potassium (K^+ aq) ions do not leach and can accumulate on the surface; soil appears white. When this occurs, soil pH may exceed 9.0 and crop plants do not grow.

Acidic soils are common where there is concentration of industry and vehicles. Combustion of coal and petroleum releases sulfur dioxide and oxides of nitrogen, which can form sulfuric and nitric acids in the atmosphere, resulting in acid rain that in turn, pollutes streams, lakes, and groundwater and can damage buildings, harm vegetation and soils, and will decrease soil pH.

Therefore it is important to calculate the amount of lime to add to achieve the desired result.

The amount needed to raise pH by a given amount (say 1 pH unit) depends on the amount of humus and clay the soil contains. Lime dissolves slowly and releases calcium and bicarbonate ions:

$$CaCO_3(s) + H_2O(l) + CO_2(g) \rightleftharpoons Ca^{2+}{}_{(aq)} + 2HCO_3^-{}_{(aq)}$$

Bicarbonate neutralizes hydronium ions, whereas the calcium ions displace hydronium and aluminum ions held by the negative charges on the surfaces of humus and clay particles. Once displaced into solution, hydronium and aluminum cations may be neutralized by the bicarbonate.

It may reduce phosphorus availability and cause deficiencies of micronutrients such as manganese (in wheat) and zinc (in pastures). Overliming may also increase the loss of sulfate if excess calcium is leached. The more humus and clay a soil contains, the greater its reserve acidity (i.e., aluminum content) and the greater amount of lime needed to raise the soil pH by a given amount. Organic and clay-rich soils containing humus and clays have considerable buffering capacity. It requires less lime to raise the pH of sandy soils.

How Much Lime Is Needed?

A reasonable question is: How much lime should be applied to achieve the necessary increase in pH? Table 3.4 offers a reasonable answer for deeper rooted agronomic crops (e.g., corn, wheat, alfalfa) and for shallow-rooted turf. The rates in Table 3.4 are guides, not recommendations. Specific recommendations can be obtained from Cooperative Extension Service personnel in each state. Their recommendations will consider the crop to be grown, liming history of the soil, the

TABLE 3.4 Approximate Tons of Lime/acre Required to Increase Soil pH to 7.0 for Alfalfa and Soybeans in New York State and Illinois (Ketterings et al., 2006; Fernandez and Hoeft, 2002)

Initial Soil pH	Sands		Silt Loams + Clays + Silty					
			Loams		Loams		Clay Loams	
	NY	IL	NY	IL	NY	IL	NY	IL
4.5	2.5	6	6	10	9.5	11	13	11
5.0–5.1	2	4.8	5	7.8	7.5	9.3	10.5	11
5.6–5.7	1	3.3	2	5.1	3	6	4.5	7.4
6.0–6.1	0.5	2.3	1.5	3.3	2	3.8	3	4.6
6.6–6.7	0.5	1	0.5	1	0.5	1	1	1

specific soil and its cation ion exchange capacity,[8] threshold pH desired, whether the lime is to be surface applied or incorporated, depth of incorporation, planned crop rotation, and annual rainfall. An additional factor is if 3–4 tons/acre is needed sequential application over 2–3 years may be agronomically and economically advisable. How much lime is needed or should be applied is often based only on a quick pH test, but the answer should consider several factors in addition to pH.

Lime rates in Table 3.4 are based on an 8-inch depth of tillage and should be increased or decreased by about 12% for each inch of shallower or deeper incorporation. The amount recommended for Illinois soils is higher because in general Illinois soils have a higher organic matter content and a higher buffering capacity because of the types of clay minerals present. Therefore, they have a higher cation exchange capacity and require more lime to achieve the desired pH. Because turf is shallow rooted and not tilled, lime rates are lower and expressed in pounds/1000 square feet. If soil pH is 5.5–6 and the desired pH is 7, sands require 20, loams 25, and clay 35 pounds/1000 square feet,[9] which equals 0.75, 1, and 1.5 tons/acre.

Pesticides are applied in pounds, fractions of a pound, or ounces per acre. Fertilizers are applied in tens of pounds per acre. Because of soil's large buffering capacity, lime is unique among soil chemical additives in that it is usually applied in tons per acre. Its cost in dollars per ton is far lower than other chemicals used in production agriculture.

AGRICULTURAL ROLE

Lime did not change agricultural practices in the same way other chemicals did. It made agricultural production possible on acidic soils where it had previously been impossible or, if possible, production was always low. Lime assured the ability of the world's farmers who must till acidic soils to fulfill their primary moral obligation: to feed people.

REFERENCES

Chandran, D., Sharopova, N., Ivas huta, S., Gantt, J.S., VandenBosch, L.A., Samac, D.A., 2008. Transcriptome profiling identified novel genes associated with aluminum toxicity, resistance and tolerance in *Medicago truncatula*. Planta 228, 151–166.

Fernandez, F.G., Hoeft, R.F., 2002. Managing soil pH and crop nutrients (Chapter 8) In: Illinois Agronomy Handbook. Department of Crop Sciences, Champaign-Urbana, IL, pp. 91–112. See: http://extension.cropsci.illinois.edu/handbook/pdfs/chapter08.pdf (accessed July 2014).

Ketterings, Q.M., Reid, W.S., Czymmek, K.J., 2006. Lime Guidelines for Field Crops in New York. Department of Crop and Soil Sciences Extension Series E06-2 Cornell University, Ithaca, NY. 35 pp. Adapted from: http://nmsp.cals.cornell.edu/publications/extension/LimeDoc2006.pdf (accessed July 2014).

---

8. Cation exchange capacity is the maximum quantity of cations a soil is capable of holding for exchange with the soil solution at a given pH value.

9. http://www.ext.colostate.edu/mg/gardennotes/222.pdf. Accessed July 2014.

Samac, D.A., Tesfaye, M., 2003. Plant improvement for tolerance to aluminum in acid soils – a review. Plant Cell, Tissue Organ Culture 75, 189–207.

Shao Jian Zheng, 2010. Crop production on acidic soils: overcoming aluminum toxicity and phosphorus deficiency. Annals of Botany 106 (1), 183–184.

Tesfaye, M., Temple, S.J., Allan, D.L., Vance, C.P., Samac, D.A., 2001. Overexpression of malate dehydrogenase in transgenic alfalfa enhances organic acid synthesis and confers tolerance to aluminum. Plant Physiology 127, 1836–1844.

Von Uexkull, H.R., Mutert, E., 1995. Global extent, development and economic impact of acid soils. Plant and Soil 171, 1–15.

Chapter 4

Nitrogen

Chapter Outline

NITROGEN AND LIFE

Nitrogen (N, atomic number 7) is a colorless, odorless gas of diatomic molecules (N_2) at room temperature.[1] It is about seventh in total abundance in the solar system and is 78% of the Earth's atmosphere, 4 quadrillion metric tons (10^{12}) is N_2. It was discovered and isolated as a separable component of air in 1772 by Scottish physician Daniel Rutherford (1749–1819), who called it noxious or fixed air.

Many industrial compounds (e.g., ammonia, nitric acid, organic nitrates (propellants and explosives), and cyanides) contain nitrogen. Large amounts of energy are released when nitrogen compounds burn, explode, or decay to nitrogen gas. Synthetically produced ammonia and the agricultural fertilizers it has enabled are essential to modern agriculture's productivity. Fertilizer nitrates are also pollutants and cause eutrophication[2] of water systems. Nitrogen is a component of Kevlar fabric, cyanoacrylate "super" glue, and most major pharmacological drugs, including antibiotics. The organic nitrates, nitroglycerin and nitroprusside, by metabolism to natural nitric oxide, which dilates arteries, act as vasodilators to control blood pressure and the onset of angina. Plant alkaloids (often defense chemicals) contain nitrogen. Nitrogen-containing drugs, such as caffeine and morphine, are either alkaloids or synthetic mimics that

---

1. Much of the following is from http://en.wikipedia.org/wiki/Nitrogen. Accessed February 2014.
2. Excessive nutrients in a body of water, which cause dense growth of plants and death of animal life from lack of oxygen.

Six Chemicals That Changed Agriculture. http://dx.doi.org/10.1016/B978-0-12-800561-3.00004-3

TABLE 4.1 Elemental Composition of the Human Body

Chemical	Percent	Atomic
Element of Mass Percent		
Oxygen	65	24
Carbon	18.5	12
Hydrogen	9.5	62
Nitrogen	3.2	1.8
Calcium	1.5	0.22
Phosphorus	1.1	0.78

act (as many plant alkaloids do) upon receptors of animal neurotransmitters (e.g., amphetamines). Industrial nitrogen gas is produced by fractional distillation of liquid air, or by mechanical means (high visible light–induced photocatalytic activity) using gaseous air. Nitrogen gas can also be produced by treating an aqueous solution of ammonium chloride with sodium nitrite.

$$NH_4Cl_{(aq)} + NaNO_{2(aq)} \Rightarrow N_{2(g)} + NaCl_{(aq)} + 2H_2O_{(l)}$$

The natural nitrogen cycle describes movement of nitrogen by all living organisms from the air to the biosphere and back to the atmosphere. Nitrogen is taken up in solution by green plants and algae as nitrate to build up the nucleotide bases needed for constructing DNA and RNA and all amino acids (a carboxylic acid with an attached amine group). Nucleotides are organic molecules composed of a nitrogenous base, a five-carbon sugar, and at least one phosphate group. Nonphotosynthetic creatures (mammals) obtain their nitrogen by consuming other living things. Soil microorganisms convert nitrogen compounds to nitrates that can be used by plants. Nitrifying bacteria live in a symbiotic relationship on the roots of legumes and "fix" molecular nitrogen directly from the atmosphere.

The human body contains about 3%, by weight, of nitrogen, the fourth most abundant element in the body (Table 4.1). We and all other living creatures cannot survive without fixed nitrogen. More than 98% of the human body is composed of the six essential elements shown in Table 4.1.

FRITZ HABER AND CARL BOSCH[3]

Fritz Haber is not a name known by many; it is highly probable that the vast majority of the Earth's people have never heard of him. Yet his most important

---

3. See Smil (2001) for a detailed presentation of Haber and Bosch, their process, and its influence on world food production.

work—the synthesis of ammonia—has had important, enduring effects on the practice of agriculture, the diet, and the well-being of all citizens of the modern world. Smil (2001) proposes that "The industrial synthesis of ammonia from nitrogen and hydrogen has been of greater fundamental importance to the modern world than the invention of the airplane, nuclear energy, space flight, or television." This is a bold, correct claim that is especially relevant to agriculture and thus to feeding the world.

Fritz Haber (1868–1934)[4] was born in Breslau, Prussia (now Wroclaw, Poland). Between 1886 and 1891, he studied at the University of Heidelberg, the University of Berlin, and the Technical College of Charlottenburg, Germany. After his early studies, he worked in his father's chemical business and in the Swiss Federal Institute of Technology in Zürich. After a few years, he moved to the Technische Hochschule's (Institute of Technology) Department of Chemical and Fuel Technology at the University of Karlsruhe, Germany (founded in 1825) where he worked as a physical chemist from 1894 to 1911. In 1905, Haber accomplished successfully what chemists had long sought to do—fix N_2 from air and combine it with hydrogen (H_2) gas to synthesize ammonia (NH_3). He used high pressure and a catalyst (iron + potassium hydroxide [KOH]). Carl Bosch,[5] a chemist and engineer employed by Badische Anilin und Soda-Fabrik (BASF) studied high pressure chemistry between 1909 and 1913. In 1910, he succeeded in transforming Haber's tabletop method of fixing nitrogen into an important industrial process that eventually produced megatons of fertilizer and much smaller amounts of explosives. The Haber–Bosch process was separated production of nitrogen products (e.g., fertilizer, explosives, chemical feedstocks) from natural nitrate deposits.

THE HABER–BOSCH PROCESS

The Haber–Bosch process is the catalytic synthesis of ammonia from the hydrogen in natural gas (CH_4) and inert atmospheric nitrogen under high temperature (400–450 °C) and pressure (200 atm = approximately 3000 psi). Natural gas is 70–90% of the cost of the process. For several years the United States imported natural gas primarily from Trinidad (45%) and Canada (23%). With recent increased production of natural gas, imports have declined and less than 3% of natural gas that supplies 24% of US energy is now imported. The methane in natural gas is converted to hydrogen in a series of simultaneous reactions:[6]

$$CH_4 + H_2O \Rightarrow 3H_2 + CO$$

4. http://en.wikipedia.org/wiki/Fritz_Haber. Accessed January 2014.
5. Carl Bosch shared the 1931 Nobel Prize in chemistry with the French chemist Friedrich Bergius "in recognition of their contributions to the invention and development of chemical high pressure methods." Bosch's development of high-pressure methods was key to the industrial success of Haber's work.
6. nzic.org.nz/chemprocess. Click on Ammonia and Urea Production. Accessed January 2014.

$$CH_4 + 2H_2O \rightleftharpoons 4H_2 + CO_2$$

$$CO + H_2O \rightleftharpoons H_2 + CO_2$$

Air mixed with the gases gives a hydrogen:nitrogen ratio of 3:1. Water, carbon dioxide, and carbon monoxide poison the required iron/KOH catalyst and are removed. Carbon monoxide is converted to carbon dioxide, which is used for urea [$CO(NH_2)_2$] production and is removed. Cooling liquifies water, which is removed. Nitrogen and hydrogen react to yield ammonia.

$$N_{2(g)} + 3H_{2(g)} \rightleftharpoons 2NH_{3(g)}$$

Initial ammonia yield is approximately 15%, but with continual recycling the yield can be 98%. Smil (2001, p. 109) points out that if Haber or Bosch were to visit a modern ammonia synthesis plant, they would be amazed by the size of the operation and the amount produced, but would recognize that the essentials of their process were the same.

The reaction had been known for many years, but yields were small and the reaction was slow. Haber and Bosch and their colleagues determined that high temperature, high pressure, and the iron oxide/KOH catalyst were necessary to increase ammonia yield. The goal was to find a way to produce ammonia for the fertilizer industry and thereby escape farming's dependence on the rapidly disappearing deposits of guano and saltpeter. Saltpeter (KNO_3), a water-soluble, white solid was one of the world's primary sources of the nitrate anion (NO_3) for fertilizer. Alexander von Humboldt (1767–1835), a Prussian geographer, naturalist, and explorer, discovered guano in Peru in 1802 and studied its fertilizing potential. His writing led to its becoming Europe's and the developed world's primary source of the nitrate anion for fertilizer, in spite of the fact that the source was thousands of miles away. Guano is a mixture of KNO_3 and $NaNO_3$. It is 8–16% nitrogen, 8–12% equivalent phosphoric acid, and 2–3% equivalent potash. It is the excrement of seabirds, cave-dwelling bats, and pinnipeds[7] (see Chapter 5 for more on guano).

In the late 1800s, Peruvian saltpeter replaced guano as the primary source of nitrogen and phosphorus fertilizer for agriculture and remained so until the Haber–Bosch method was developed in 1909. Peru and Bolivia both had large reserves of high-quality nitrate and phosphorus-rich deposits of guano. Guano's newfound value as a fertilizer and saltpeter's role as a fertilizer and for manufacture of explosives made the Atacama Desert (Northern Chile) strategically and economically important. Bolivia, Chile, and Peru were located in the area of the largest reserves of a resource the world demanded. After the Saltpeter War (1879–1883) (also known as the Guano War, the Guano and Saltpeter War, and the Second War of the Pacific), Bolivia lost its territory, which had extended its boundary to the

---

7. A pinniped is an aquatic carnivorous mammal (e.g., seal, walrus) with all four limbs modified into flippers.

Pacific coast and the guano and saltpeter reserves in that area. Peru held 59% of all saltpeter reserves, and Chile had 19% for most of the nineteenth century. As is true for any natural resource with a finite supply, guano was depleted rapidly. Ammonia and nitrates were also produced from the destructive distillation of coal, but the quantity was grossly insufficient to satisfy agriculture's need. "By 1900, the US used 48% of its imports of $NaNO_3$ to make explosives" (Smil, 2001, p. 47).

Poison Gas and World War I

The Haber–Bosch process also enabled production of explosives, which many might not regard as a similarly positive contribution to humanity. Le Couteur and Burreson (2003) provide a complete description of the role of nitrogen and nitro compounds in the development of explosives including gunpowder, ammonium nitrate, and pentaerythritol tetranitrate. The latter is the base of plastic explosives and their uses in mining and war: a subject beyond the scope of this book.

Haber has also been described as the father of chemical warfare for research that led to the development and deployment of a new weapon: chlorine and other poisonous gases, during World War I, a singularly negative achievement. During the war, Haber threw his energies and those of his institute into support of the German war effort. We cannot know for sure, but it is generally accepted that his development of and promotion of the use of poisonous gas led to his wife's suicide on the night of his promotion to captain in the German army. Many others condemned him for his wartime role. In addition to leading teams to develop chlorine and other deadly gases for use in trench warfare, Haber supervised initial deployment of chlorine gas on the Western Front at Ypres, Belgium, in 1915. He participated in its release at the Front during the war, despite its proscription by the Hague Convention of 1907 (to which Germany was a signatory). Haber said, "During peace time a scientist belongs to the world, but during war time he belongs to his country." He surely did not regard his position as an ethical dilemma. Many did.

Poison gas was used primarily to demoralize, injure, and kill entrenched defenders. The gases included disabling chemicals (tear gas: a group of synthetic organic halogen compounds), the more severe mustard gas (sulfur mustard [$(Cl–CH_2CH_2)_2S$]) to lethal gases such as phosgene ($COCl_2$) and chlorine (Cl_2). The British also used poison gas in World War I. Perhaps they invoked what one might call the Brass rule: Do unto others as they have done unto you!

Haber's research, advocacy, and use of chemical warfare killed less than 89,000 people (less than 0.01% of the 10 million military deaths). Another 1.2 million soldiers had nonfatal, serious, often enduring effects. He was a proud recipient of the Iron Cross for his work on gas warfare. During the 1920s, scientists working at Haber's institute, the Kaiser Wilhelm Institute for Physical Chemistry and Electrochemistry in Berlin, developed methylcyanoformate (a cyanide gas formulation) called Zyklon A ($C_3H_3NO_2$). It was used as an insecticide and a fumigant in grain stores. Zyklon B (hydrogen cyanide gas + a stabilizer, odorant, and adsorbents) was used by the Nazis to kill 1.2 million people, including 960,000 Jews, during World War II.

The Nobel Prize

Haber was awarded the 1918 Nobel Prize in chemistry "for the synthesis of ammonia from its elements." The result of his research is the industrial manufacture of nitrogen fertilizer that enables food production for as much as half of the world's population of 7.2 billion people. The Haber–Bosch process consumes more than 1% of human energy production. On average, half of the nitrogen in a human body comes from synthetically fixed sources, the product of a Haber–Bosch plant.

The great consternation when Haber was awarded the Nobel Prize in chemistry for the synthesis of ammonia stemmed from the fact that he had been an enthusiastic advocate of chemical warfare during World War I and directed German research that resulted in thousands of deaths. There have been posthumous calls to strip Haber of his Nobel Prize. There's something oddly appropriate, yet ironic, about Haber's award, however, because the award was funded, as all Nobel Prizes are, by Alfred Nobel, a German chemist, who held the original patent (1867) for dynamite. Haber did not receive the prize in 1919 as was customary, but in 1920. No members of the Swedish Royal family were present.

The Chemistry of the Haber–Bosch Process

The chemistry and techniques for the effective synthesis of ammonia did not spread to the rest of the world until the 1920s and 1930s. The Haber–Bosch process was a milestone in industrial chemistry because it divorced production of nitrogen fertilizer, explosives, and chemical feedstocks from natural guano and saltpeter.

From the 1840s to the present day, various deposits of phosphate rocks and potash have provided adequate sources of the required fertilizer elements: phosphorus (see Chapter 5) and potassium. The process and techniques of fertilizer preparation and production have changed but the chemical reactions and principles remain basically the same. This has not been true for nitrogen. The Earth's atmosphere is the only source.

Nitrogen fertilizer is produced by conversion of ammonia to nitric acid in three stages. The first stage creates nitrous oxide from ammonia and oxygen.

$$4NH_{3(g)} + 5O_{2(g)} \Rightarrow 4NO_{(g)} + 6H_2O_{(g)}$$

In stage two, carried out in the presence of water, nitric oxide is oxidized again to yield nitrogen dioxide.

$$2NO_{(g)} + O_{2(g)} \Rightarrow 2NO_{2(g)}$$

Stage three is the ready solubility of nitrogen dioxide in water, which yields nitric acid.

$$3NO_{2(g)} + H_2O \Rightarrow 2HNO_{3(aq)} + NO_{(g)}$$

Nitric acid is reacted with ammonia to manufacture agricultural fertilizers (e.g., ammonium nitrate), which because of its explosive potential is no longer available.[8]

$$HNO_{3(aq)} + NH_3 \Rightarrow NH_4NO_{3(aq)}$$

Other explosives such as dynamite, which is nitroglycerin ($C_3H_5N_3O_9$) adsorbed to diatomaceous earth or another absorbent substance, as a carrier for nitroglycerin. Nitric acid was also used to manufacture trinitrotoluene.

Ammonia created by the Haber–Bosch process can be injected directly into any nitrate-containing fertilizer. For example, the ammonia will react with potassium nitrate or other nitrates present to form ammonium nitrate. Anhydrous ammonia can be injected directly into an organic material such as wheat straw to form urea an artificial protein and nitrogen fertilizer. Urea—carbamide [$CO(NH_2)_2$]—plays an important role in the metabolism of nitrogen-containing compounds by animals. It is the nitrogen-containing substance in animal urine.

Urea can be synthesized from ammonia and carbon dioxide in a two-step process under high pressure and temperature.

$$2NH_3 + CO_2 \Rightarrow COO(NH_2)_2 \text{ (ammonium carbamate)}$$

As pressure is reduced and heat increased, ammonium carbamate decomposes to ammonia and carbon dioxide.

$$COO(NH_2)_2 \Rightarrow CO(NH_2)_2 \text{ (urea)} + H_2O$$

The urea solution is concentrated to nearly 100% urea and granulated for fertilizer.

In 1828, the German chemist Friedrich Wöhler produced urea from inorganic materials by treating silver cyanate with ammonium chloride.

$$AgNCO + NH_4Cl \Rightarrow [CO(NH_2)_2] + AgCl$$

This was the first time an organic compound was artificially synthesized from inorganic starting materials without the involvement of living organisms or other biological material. His work was an important conceptual milestone in chemistry. Wöhler's work disputed and essentially falsified the doctrine of vitalism, which held that living organisms are fundamentally different from nonliving things because they contain a nonphysical element and are governed by different principles than inanimate things. It was the prevailing theory, but the core element of all scientific hypotheses is that they must be capable of being proven false (the Principle of Falsifiability), Wöhler's work falsified the doctrine of vitalism, which was rejected.

---

8. The A.P. Murrah Federal Building in Oklahoma City was destroyed on April 19, 1995, by a planned explosion of ammonium nitrate fertilizer. The bombing killed 168 people, injured more than 680, destroyed or damaged 324 buildings, destroyed 86 cars, shattered glass in 258 buildings, and caused $652 million worth of damage.

NITROGEN FIXATION

Natural Nitrogen Fixation

In the early twentieth century, there were four natural and three synthetic methods to "fix" nitrogen, that is, to convert inert, inactive atmospheric nitrogen gas to nitrogen compounds (primarily ammonia) that could be used to produce fertilizer. Nitrogen fixation can also refer to other biological actions such as conversion to nitrogen dioxide. The natural methods of nitrogen fixation are as follows.[9]

Lightning. Fixation of atmospheric nitrogen by lightning was regarded as an important source of soil nitrogen.

Microorganisms that fix nitrogen are prokaryotes. All bacteria and archaea that fix atmospheric nitrogen gas into a more usable form such as ammonia are diazotrophs. They are prokaryotic not eukaryotic cells which contain membrane-bound organelles (e.g., nucleus, mitochondria). Biological nitrogen fixation was discovered by the German agronomist H. Hellriegel and the Dutch microbiologist M. Beijerinck in 1901. It is enzymatic conversion of atmospheric nitrogen to ammonia facilitated by the nitrogenase enzyme. The reaction is

$$N_2 + 8H \Rightarrow 2NH_3 + H_2$$

Five genera of microorganisms can fix atmospheric nitrogen: Cyanobacteria (dominantly *Trichodesmium*), green sulfur bacteria, Azotobacteraceae, Rhizobia, and Frankia.

Symbiotic organisms. Nine genera (not all species in each genera) of higher plants, and some insects (e.g. termites), have formed symbiotic associations with diazotrophs that enable nitrogen fixation. Lichens illustrate a symbiotic relationship between an algae and a fungus.

Legumes. Most members of the *Leguminosae* (e.g., alfalfa, bean, clover, peanut, pea, lentil, soybean) contain symbiotic Rhizobia bacteria within nodules on their roots, which produce nitrogen compounds that help the plant grow and compete with other plants. When the plant dies, the fixed nitrogen is released, making it available to other plants. This is an important reason legumes are desirable rotational crops; they provide nitrogen. The wide variety of nitrogen fixing legumes, found in diverse ecosystems and climates, share two traits:

- They reproduce by seed borne in a seed pod.
- They have symbiotic nitrogen-fixing bacteria in root nodules.

The symbiotic relationship enables direct nitrogen fixation and eliminates the need for nitrogen fertilization to improve yield. However, the few crops (wheat, rice, and corn) that provide most of the food for humans from terrestrial ecosystems cannot fix atmospheric nitrogen.

---

9. http://en.wikipedia.org/wiki/Nitrogen_fixation.

Synthetic Nitrogen Fixation

Two methods—the electric arc process and the cyanamid process[10]—were known and used, but have never been important for fertilizer production. The cyanamid process is used to manufacture the fertilizer calcium cyanamide. It requires temperatures of 1000–1100 °C, usually with a catalyst in a mixture of nitrogen and calcium carbide, which form calcium cyanamide and carbon:

$$CaC_2 + N_2 \Rightarrow CaCN_2 + C$$

Calcium cyanamide reacts with water (steam) to create calcium carbonate and ammonia:

$$CaCN_2 + 3H_2O \Rightarrow CaCO_3 + 2NH_3$$

The second phase can proceed in soil where ammonia is oxidized to nitrate and used by plants.

The electric arc is a device in which an electric current (a flow of electrons) flows between a cathode and an anode, separated by a gas (lightning is an electric arc). Electric arc refers to the device and the electric discharge. It is capable of combining free atmospheric nitrogen with hydrogen to form inorganic compounds, such as ammonium ions, which can then be converted by nitrification into nutrients that can be absorbed by plants.

At the beginning of the twentieth century, the British chemist Sir William Crookes delivered a lecture—The Wheat Problem—to the British Association for the Advancement of Science in Bristol. He said the wheat crop was dependent on nitrogen fertilizer obtained from Chilean saltpeter, which was being rapidly depleted. When this was combined with the fact that Europe had no new land on which more wheat could be grown, Crookes concluded that another source of fertilizer nitrogen was urgently needed.

The third way to fix/synthesize atmospheric nitrogen, the Haber–Bosch method, solved the problem Crookes identified. It has made a huge difference to agriculture and the world. Although the point can be debated, it is reasonable, within the agricultural domain, indeed for the world, to regard the Haber–Bosch process as the most important technological advance of the twentieth century, as Smil (2001) has. Enabling production of nitrogen fertilizer, essentially from the air, must be regarded as a positive achievement because it improved worldwide crop yields. It, more than any other scientific achievement, allowed farmers and agricultural scientists to fulfill their primary moral obligation to produce food.

Because of World War I and wartime secrecy, the method became a German national secret. BASF opened the first ammonia synthesis plant to produce nitrogen fertilizers (primarily ammonium sulfate) in Oppau, Germany, in 1913.[11]

---

10. See http://www.scienceclarified.com/Di-El/Electric-Arc.html#ixzz2ncshoumL. Accessed December 2013.

11. http://www.basf.com/group/corporate/us/en/about-basf/history/1902-1924/index. Accessed January 2014.

The Oppau plant exploded in 1921, killing more than 500 people and destroying more than 7000 homes. It was the largest catastrophe in German industrial history. The annual output of 7200 metric tons of ammonia produced 36,000 metric tons of ammonium sulfate. Its initial purpose was to secure food supplies for a growing population. By the end of 1917, Germany had two ammonia plants that produced 90,000 Mt of nitrates. A shortage of ammunition loomed and nearly all production had to be delivered to the explosives industry, not to agriculture. German chemical expertise and its synthetic chemical industry enabled its entry and near success in two world wars. That industry began with a search for quinine substitutes, which is reported in more detail in Chapter 7. One is compelled to wonder if German chemists' expertise and early success in developing the chemistry of explosives, dyes, fertilizers, pharmaceuticals, and synthetic petroleum enabled Weimar Germany to lead the world—in chemistry and into war.

RESULTS OF THE HABER–BOSCH PROCESS

Positive Results

The process is generally credited with enabling Germany to manufacture munitions and maintain crop production during World War I, after the British naval blockade cut off supplies of nitrates from Peru and Chile. It was obvious to all agricultural scientists that nitrogen fertilizer increased crop yield. As early as the late nineteenth century, some scientists were concerned about rapid depletion of natural sources of nitrogen that could be used as fertilizer.

The more important result of the Haber–Bosch process is that the food supply of half of the current world population is dependent on nitrogen fertilizer produced by the Haber–Bosch process. There are numerous large-scale ammonia production plants worldwide, producing a total of 198 million Mt in 2012.[12] Four countries produce half: China 28.6%, India 8.6%, Russia 8.4%, and the United States 8.2%. Eighteen countries produce at least 1000 Mt annually. Twelve companies had 24 plants in 16 US states in 2010. Sixty percent of total US ammonia production was in Louisiana, Oklahoma, and Texas because of their large reserves of natural gas. The global demand for nitrogen fertilizer in 2013 was approximately 109 million metric tons,[13] 13% of which is consumed in North America. Eighty percent is used by agriculture. Urea is the dominant form because ammonium nitrate use has been banned (see footnote 9). About 69% of US cropland planted with major, nonleguminous field crops (barley, corn, cotton, oats, potato, sorghum, and wheat) receives commercial nitrogen (commonly as a complete fertilizer with N, P, and K). Corn is 45% of US crop acreage—80 million major acres—and is the major user of nitrogen fertilizer.

---

12. http://en.wikipedia.org/wiki/Ammonia. Accessed November 2104.
13. http://www.ers.usda.gov/amber-waves/2011-september/nitrogen-footprint.aspx#.UtwAZhDn_b0. Accessed January 2014.

It receives 65% of the 8.7 million tons of nitrogen applied by farmers each year, which in 2013 produced 13.99 billion bushels (7.0 million tons) of corn. The United States produces 32% of the world's corn. Livestock feed consumes 46% of the US crop, 22% is used to produce ethanol, 20% is exported, and 12% is used to make high fructose corn syrup starch and other sweeteners.

Corn production/acre averaged 54.7 bushels in 1960. The 2013 crop (13.99 billion bushels) was produced on 97.4 million acres. To produce the 2013 crop with 1960 yields would require an additional 2.4 billion acres (2.6 times more land). The land is not available. It does not exist. The world has 3.7 billion acres of cropland, nearly all of which is used to grow crops. The truly spectacular yield increases must be credited to the contributions of agricultural science, improved production technology, and the skill of farmers, but without the nitrogen fertilizer contributed by the Haber–Bosch process, the other contributions could not have almost tripled corn yield since 1960.

Negative Results

- **A human cost**
  One negative result of the Haber–Bosch process was the collapse of Peruvian/Chilean nitrate market and production. It fell from 2.5 million tons selling for $45.00/ton and employing 60,000 workers in 1925 to 800,000 tons, produced by 14,133 workers, and selling for $19.00/ton in 1934.
- **Monocrop agriculture**
  The world's farmers are rational, intelligent businessmen and women. They are engaged in a demanding task to produce more and more food for a growing population with greater demands. The ready availability of fertilizers, particularly nitrogen, enabled farmers to produce the spectacular yield increases of crops that yielded the greatest monetary return. Monocultural agriculture is a result. Many farmers grow one or two primary crops and, although they may rotate the fields on which crops are grown, they grow the same ones every year. Rotation to legumes or long-term pasture or hay crops has not disappeared, but it has become much less common. Although the change is economically rational, it is environmentally irrational and, probably inevitably, an agricultural mistake. The mistake results in reduced soil tilth, increased erosion, reduced water-holding capacity, and a loss of soil microbial diversity (Smil, 2001). Fertilizers and pesticides have made this possible. Many argue that it is a sustainable system. Others, with equal conviction, claim that, in the long run, it is not sustainable. No one knows how long the long run may be.
- **Pollution**
  As is true for many major scientific achievements, not all the results have been beneficial. There is a caveat. The intensive (amount/acre) and extensive (1960: 71.4; 2013: 97.4 million acres) use of nitrogen fertilizer makes agriculture the single largest source of nitrogen compounds entering the environment in the United States, contributing 73% of nitrous oxide emissions, 84% of

ammonia emissions, and 54% of nitrate emissions in recent years (Ribaudo 2011). Only about 35% of US crop acres that receive nitrogen annually meet all nitrogen management criteria, leaving 65%, more than 100 million acres, that need improved management (Ribaudo). Smil (2001, p. 178) suggests that "the Haber-Bosch synthesis introduces about four times as much reactive nitrogen into the biosphere as does the combustion of fossil fuels." More than half of all the nitrogen fertilizer used in all of human history has been used since 1990 (Clayton, 2004). Half of the nitrogen fertilizer used each year on agricultural crops is not used by the crops, it is lost to the environment and much of it ends up in the atmosphere or local waterways, releasing 2.1 billion tons of carbon dioxide equivalent as nitrous oxide, which can stay in the atmosphere as long as 120 years and has about 300 times greater effect than carbon dioxide on global warming (WorldWatch, 2008)—a potent greenhouse gas. Fifty to 60% of the 1.5 billion pounds of nitrogen fertilizer applied primarily to US cornfields each year is used by the corn; the rest is free in the environment.

It is also true that animal agriculture has become more concentrated. There are more than 250,000 confined animal feeding operations—concentrated animal feeding operations in the United States (see Chapter 9) inevitably produce manure and liquid waste that have high concentrations of nitrogen, other nutrients, and heavy metals. Pollution is an undesirable result.

Eutrophication is a process by which a body of water (e.g., a lake) becomes gradually enriched in the concentration of phosphates, nitrogen, and other plant nutrients that stimulate the growth of aquatic plant life, usually resulting in the depletion of dissolved oxygen. It is commonly a result an abundant supply of nitrogen from fertilizer and sewage effluent (Tilman et al., 2001). As the concentration of dissolved nutrients in the ecosystem increases, the amount of organic material that can be broken down to available nutrients increases. The nutrients released effectively become pollutants. Undegraded organic material enters the ecosystem mainly through runoff.

Algal blooms often develop on the water surface, preventing light penetration and oxygen absorption necessary for underwater life. A hypoxic (deprived of adequate oxygen) zone is created. Nitrogen fertilizer has contributed to the hypoxic (O_2 concentration <2 mg/L) zone in bottom water on the Louisiana–Texas coast. The sediment load in the Missouri/Mississippi river basin is about 616 million tons[14] (550 million metric tons) annually. Much of the nitrogen that does not fertilize corn and other crops grown in the Mississippi/Missouri river system reaches the Gulf of Mexico in sediment and is agriculture's contribution to the hypoxic zone.

The ultimate effect of excess nutrients is an algal bloom. Dead algae sink to the bottom, and oxygen is used for decomposition at a rate faster than it can be added

---

14. A metric ton is a unit of mass equal to 1000 kg. It is approximately equal to 2204.6 pounds or 1.10 US tons (2000 pounds).

back into the system by physical mixing. The lack of oxygen (anoxia) kills bottom-feeding organisms (e.g., crabs, shrimp, catfish) and creates a dead zone, which affects the work and income of those whose living depends on catching them.

The US area begins where the Mississippi River enters the Gulf of Mexico. In 2010, it extended east to Alabama and west to Galveston, TX. The area in mid-2010—7722 square miles—was 11% less than the area of New Jersey (8722 square miles). In 2014, the area had decreased to 5052 square miles (8% smaller than Connecticut, 5543 square miles) because of mixed conditions on the southeastern part of the study area and winds from the west that pushed the hypoxic water mass east, thus reducing the bottom area footprint.

There is a growing consensus that corn grown for ethanol production in the United States exacerbates the problem because of high nitrogen fertilizer use and the substitution of corn for soybeans, which do not require nitrogen fertilizer. The combination of increasing corn acreage, nitrogen fertilizer, the quest for ever-higher production, and government subsidies for ethanol production are primary causes of Gulf of Mexico hypoxia (Goolsby et al., 2001; Rabalais et al., 2002). Mean annual nitrate N concentrations at St. Francisville, LA, from 1980 to 1996 were more than double the average concentration from 1955 to 1970 (Goolsby et al.).

Hypoxia is not limited to the United States. It has spread rapidly since the 1960s. Worldwide, there are now more than 500 dead zones covering 250,000 km^2 (96,525 square miles; 1% smaller than Wyoming, 97,813 square miles) with the number doubling every 10 years since the 1960s. Postel reported 146 areas in the world in 2005 (p. 23), some as small as a square kilometer, others up to 27,000 square miles. Diaz and Rosenberg (2008) reported 405 areas in 2008. The economic costs to fisheries, tourism, and other coastal livelihoods are already in the many tens of billions of dollars annually and will only continue to increase in a business as usual scenario.[15]

The largest hypoxic dead zone is in the Baltic Sea (northern Europe), home to seven of the world's 10 largest marine dead zones. Ten countries share the Baltic Sea area. The primary crops are cereals (e.g., wheat) and rapeseed (canola). Since the mid-1990s, annual inputs of nitrogen have steadily decreased. Agricultural use remains the major contributor, but human sewage effluent, animal manure, and industrial pollution are contributors (Smil, 2001, p. 192). The Gulf of Mexico is 3.9 times larger than the Baltic, but over 30 years, the latter's hypoxic area has been almost equal (18,919 square miles) to the area of Vermont plus New Hampshire (18,965.5 square miles). "Currently, hypoxia and anoxia are among the most widespread deleterious anthropogenic influences on estuarine and marine environments, and now rank with over-fishing, habitat loss, and algal blooms as major global environmental problems" (Diaz and Rosenberg, 2008). As agriculture expands its productive capability, it is inevitable that

---

15. http://www.undp.org/content/undp/en/home/librarypage/environment-energy/water_governance/ocean_and_coastalareagovernance/issue-brief—ocean-hypoxia–dead-zones-/. Accessed January 2014.

nutrient inputs will increase and hypoxia will become more widespread. More than half of all accessible freshwater is used by humans (Postel, 2005), most to irrigate crops, and hypoxia because of runoff of nitrogen and other fertilizer nutrients is inevitable. Tilman et al. (2001) forecast that the next 50 years may be last period of rapid expansion of agricultural productivity. The wealthy will demand more, especially more meat, and 2–3 billion more people will require and demand food. Agriculture will thereby continue to be a major driver of global environmental change. If agricultural practice continues to externalize its environmental costs and population and consumption continue to grow, Tilman et al. (2001) report that as much as a billion acres of natural ecosystems may be converted to agriculture by 2050. This will, inevitably, be accompanied by increases in "nitrogen, phosphorus and pesticide driven eutrophication of aquatic systems." The world faces, as it has for some time, what Kolbert (2013) identifies as the fertilizer, fertility, and population growth dilemma.

- **Overweight and obesity**
  A second, surely debatable, undesirable result of the Haber–Bosch process is a consequence of success in producing abundant food. Its very abundance has resulted in the growth of overweight and obese people. The United States produces enough food now to provide every American with 3800 calories/day. A daily caloric intake of 2350 is a healthy diet if it includes less meat than we now eat and more fruits and vegetables. Because food is available, we eat too much. There are more overweight or obese adults in developing countries, 904 million, than in the developed world, about 557 million. Similarly, more than 30 million overweight children live in the developing world compared to just 10 million in developed countries. One of every five people in the world is obese (30% heavier than the ideal body weight) and 27% of Americans are. The number of underweight (body mass index[16] below 18.5) people in the world has declined slightly since 1980 to about 1.1 billion, whereas the number of overweight (body mass index greater than 25) people is now about the same (WorldWatch, 2000). In the United States in 2009–2010, 35.7% of adults, 18.4% of adolescents 12–19 years old, 18% of children 6–11, and 12.1% of children aged 2–5 were obese.[17] There was no significant difference in prevalence between men and women at any age. Overall, adults aged 60 and older were more likely to be obese than younger adults. Among men, there was no significant difference in obesity prevalence by age. Among women, however, 42.3% of those 60 and older were obese compared with 31.9% of women aged 20–39 (Ogden et al., 2012). Obesity is a major human health problem and will consume a larger percentage of future health-care costs.

---

16. Body mass index = weight (pounds)/height (inches)^2. It is a widely used, controversial estimate of weight problems.
17. http://www.cdc.gov/nchs/data/hus/hus12.pdf#063. Accessed January 2014.

As mentioned previously, the food supply of many, no one knows how many, of the more than 5 billion people who live in what, for lack of a better term, we call developing countries depends on many factors, but the Haber–Bosch process is dominant. These are the countries with a high percentage of young people and a high birth rate, where 97% of anticipated growth of the world's population will occur. They are countries with large populations, high birth rates, a young population that has not yet begun to reproduce, and poor agricultural technology.

- **Food waste**
  Although there is no reason to suspect that it is a uniquely American trait, we do have a penchant to avoid accepting personal responsibility until all other options have been explored. There is no question that Haber and Bosch deserve some ultimate blame for the synthesis of ammonia that provided the nitrogen fertilizer that enabled exponential increases in food production. However, one must acknowledge two facts:
1. Doubling food production by 2050 will require increasing nitrogen application 2–2.5 times, which will enable more food production and exacerbate nitrogen's well-documented negative effects (Myers, 2009, p. 24).
2. The irony that the world already grows enough food to feed 10 billion people, but we can't end hunger (Holt-Giménez et al., 2012). Hunger, many argue is not a production problem, it is a poverty, inequality, and distribution problem.

As mentioned previously, the productivity of modern agricultural systems has resulted from agricultural science, technology, and the skill of farmers, but as mentioned previously, without the nitrogen fertilizer contributed by the Haber–Bosch process it would not have been possible. At least 40% (in 2010, 1.3 billion pounds) of food produced in the United States is never eaten (Gunders, 2012). A waste of $165 billion a year, which negatively affects water resources and greenhouse gas emissions. The London-based Institution of Mechanical Engineers (Fox, 2013) reported that the world's farmers produced about 4 billion metric tons of food. Because of poor harvesting, inadequate storage and transportation, and market and consumer waste, 30–50% (1.2–2 billion tons) never reaches a human stomach. Fox notes the data do not reflect the fact that large amounts of land, energy, fertilizer, and water are also wasted.

Moving food from farm to fork uses 10% of the total US energy budget, 50% of the land, and consumes 80% of all freshwater. No business would survive if 40% of their production was wasted, but that is a result of the US food production system. Surely it is rational to argue that no business that allows the equivalent of $165 billion of its product to be thrown away each year is sustainable. This is especially true when that waste is the largest component of US municipal solid waste and accounts for a large portion of US methane emissions. Fifteen percent of what is wasted would feed more than 25 million people every year. When one in six Americans and 14.5% of households (49 million people) lack a secure food supply, hunger becomes a moral problem. Increasing the efficiency of the agricultural enterprise is a way to address that problem.

It is beyond the scope of this book to prescribe how this is to be done (see Smil, 2001, pp. 217–221). It is reasonable to place some of the blame on the genius of Fritz Haber and Carl Bosch, but, although they may bear ultimate responsibility, for the human costs, pollution, overweight and obesity, and waste resulting from nitrogen fertilizer, these are our collective problems, not theirs. Accepting personal responsibility is the right choice.

EFFECTS ON AGRICULTURE

It is not a theory; it is an unquestioned scientific fact that all plants require nitrogen to grow. Successful growth of the nonleguminous crops (corn, rice, wheat, barley, cassava, millet, sorghum, soybean, and potato) that feed us requires fixed nitrogen, which can only be provided as fertilizer (Bontz, 2013). To date no substitute for the Haber–Bosch process required to make agricultural fertilizer has been developed (Bontz). In spite of decades of research on nitrogen fixation by legumes, that ability has not been transferred to nonleguminous crops. No bio-engineered technique to make nonlegumes fix nitrogen is on the horizon. Agriculture's practitioners—farmers—cannot wait for scientists to make food crops fix nitrogen as legumes do. They need to fertilize crops every year. Nitrogen fertilizer is and for the foreseeable future will continue to be essential. The research by Haber and Bosch has made it possible to produce yields that only a few decades ago were just dreams. It is a bold, but arguably accurate claim that the Haber–Bosch process has contributed more to mankind's food supply than all other achievements of chemical research.

The Nobel Prize Web site lists 10 laureates whose work changed the world; two were chemists (Marie Curie, who won a second Nobel in Physics, and Linus Pauling, who also won the Nobel Peace Prize). One hundred and nine years after their work on, what should be regarded as one of the most important achievements of chemical research, Haber and Bosch, both Nobel laureates, are not listed. If one thinks about it, which most people do not, what they achieved, affects nearly all of the 1.3 billion people who live in the world's rich countries every day. Nearly all of them have access to an abundance of food, the production of which is dependent upon Haber and Bosch's research. Their work also affects the food supply of the other 5.9 billion inhabitants of our planet.

At the end of October 2014, the world had 7.27 billion people. The population is growing at 1.14%/year, which, if constant, means the population will double in 61 years. At least 80% (more 5 3/4 billion) now live on less than US $10.00/day. There are poor people with inadequate food in all developed countries. Agricultural technological developments (fertilizers, pesticides, improved crop cultivars, irrigation, etc.) have made it possible to produce enough food to provide an adequate daily diet (about 2500 calories) for all people. However, all are not fed. Further discussion of the fact that more than 5 billion people live on less than US $10.00/day and more than 1 billion of them live on less than US $1.00/day is beyond the scope of this book. It is not a result of Haber and Bosch's work. It is a rational policy decision that reflects the fact that there is no accepted, universal right to food. No one has a duty to feed another.

An important, environmentally positive result of the widespread use of nitrogen fertilizer is the recognition that because of nitrogen's negative effects, changing the way agriculture is practiced is an increasingly important part of agricultural research. The societal benefits of nitrogen fertilizer are indisputable. However, as is true of most technological advances, there have been undesirable outcomes. The negative effects discussed previously include the increasing dominance of large-scale farms that practice monocrop agriculture. It is not inevitable, but it is clear that monocrop agriculture can and frequently has resulted in loss of ecological diversity, diminishment of soil quality, increasing farm size, and loss of small farms and rural communities. Continued use and overuse of plentiful nitrogen fertilizer has inevitably led to pollution, development of eutrophication, and hypoxic zones.

It is clearly stretching the truth and applying blame where there is no fault to suggest that Haber and Bosch are somehow responsible for the number of overweight and obese people or the waste of food. The detrimental effects of excessive weight and obesity are not due to scientific research. They are due to improper diets enabled by the abundant food available, particularly in developed countries. Many claim that the many fast-food purveyors (you know the names) and consumption of high fructose corn syrup in soft drinks should accept some responsibility. Many disagree.

In view of the enormous advances in agricultural technology, the truly amazing level of food production and the bright future for further technological developments, what should a farmer who wants to farm again tomorrow do? What are the options? Should we, can we, develop a new agricultural production system that enables farmers to take full advantage of available agricultural technology, including nitrogen fertilizer and simultaneously reduce or eliminate the negative effects of present technology? Can farmers and the agricultural system continue to feed a growing world population? Smil (2001) offers several feasible means to retain, if not increase, profitability and move toward sustainability:

- Use regular soil testing to determine the appropriate level of fertilization.
- Apply fertilizer at the proper time and, insofar as possible, place the fertilizer close to plant roots so it will be absorbed readily, but not be close enough to harm the plants. Agricultural scientists have produced abundant research about proper fertilizer placement.
- Include crop rotations with legumes.
- Reduce tillage.
- Avoid overirrigation and excessive use of water.

These are not radical recommendations. They are achievable goals that will allow continued use of nitrogen fertilizer and other agricultural technologies so important to feeding the world.

REFERENCES

Bontz, S., 2013. N'- How Humans Took a Nutrient and Their Population to a New Power, and How a New Crop Population Eased the Economic and Ecological Costs. The Land Institute. 107 (Fall 2013), pp. 11–24.

Clayton, M., May 27, 2004. 'Dead Zones' Threaten Fisheries. Christian Science Monitor. pp. 13, 16.

Diaz, R.J., Rosenberg, R., 2008. Spreading dead zones and consequences for marine ecosystems. Science 321 (5891), 926–929.

Fox, T., 2013. Global Food Waste Not, Want Not. Institute of Mechanical Engineers, London, UK. 36 pp.

Goolsby, D.A., Battaglin, W.A., Aulenbach, B.T., Hooper, R.P., 2001. Nitrogen input to the Gulf of Mexico. Journal of Environmental Quality 30, 329–336.

Gunders, D., August 2012. How America Is Losing up to 40% of Its Food from Farm to Fork to Landfill. Natural Resources Defense Council, New York. IP:12-06B.

Holt-Giménez, E., 2012. We already grow enough food for 10 billion people.... And still can't end hunger. Editorial – Journal of Sustainable Agriculture 36, 595–598.

Kolbert, E., October 21, 2013. Head Count: Fertilizer, Fertility, and the Clashes over Population Growth. The New Yorker. pp. 96–99.

LeCouteur, P., Burreson, J., 2003. Napoleon's Buttons- 17 Molecules that Changed History. Jeremy P Tarcher/Penguin, New York. 375 pp.

Myers, S.S., 2009. Global Environmental Change: The Threat to Human Health. Worldwatch report no. 181. Worldwatch Institute, Washington, DC. 48 pp.

Ogden, C.L., Carroll, M.D., Kit, B.K., Flegal, K.M., 2012. Prevalence of Obesity in the United States, 2009–2010. NCHS Data Brief, No 82. National Center for Health Statistics, Hyattsville, MD.

Postel, S., 2005. Liquid Assets: The Critical Need to Safeguard Freshwater Ecosystems. World-watch Paper 170. Worldwatch Institute, Washington, DC. 78 pp.

Rabalais, N.N., Turner, R.E., Scavia, D., 2002. Beyond science into policy: Gulf of Mexico hypoxia and the Mississippi river. BioScience 52 (2), 129–142.

Ribaudo, M., September 2011. Reducing Agriculture's Nitrogen Footprint: Are New Policy Approaches Needed? Amber Waves. U.S. Dept. of Agriculture. Economic Research Service, Washington, DC. http://www.ers.usda.gov/amber-waves/2011-september/nitrogen-footprint.aspx#.U9et3fldWzm (accessed July 2014).

Smil, V., 2001. Enriching the Earth: Fritz Haber, Carl Bosch, and the Transformation of World Food Production. The MIT Press, Cambridge, MA. 338 pp.

Tilman, D., Fargione, J., Qolff, B., D'Antonio, C., Dobson, A., Howarth, R., Schindler, D., Schlesinger, W.H., Simberloff, D., Swackhamar, D., 2001. Forecasting agriculturally driven environmental change. Science 292, 281–292.

WorldWatch, March 4, 2000. Chronic Hunger and Obesity Epidemic Eroding Global Progress. News Release. http://web.archive.org/web/20030316051456/http://worldwatch.org/press/news/2000/03/04/ (accessed November 2014).

World Watch, May/June 2008. Adjustments in Agriculture May Help Mitigate Global Warming. p. 4.

Chapter 5

Phosphorus

Chapter Outline

INTRODUCTION

Phosphorus is essential to life primarily because of its role in the transfer of energy via adenosine triphosphate (ATP), the universal energy source in all living systems. When ATP reacts with water (hydrolysis) and loses a phosphate ion, energy is released. That energy drives the innumerable processes that make life possible in all creatures. Agriculture is the intentional use of living systems to produce food and fiber. We, and the modern agricultural system that produces our food, are absolutely dependent on phosphorus. It is essential. There are no substitutes. Future agriculture will collapse without phosphorus.

Phosphorus is a highly reactive, poisonous, nonmetallic element, which burns in air, glows in the dark, and occurs naturally in phosphates, especially apatite—$[Ca_5(PO_4)_3]$—which occurs as fluoro $(F)_2$, chloro $(Cl)_2$, or, most commonly, as hydroxyapatite. Elemental phosphorus is not isolated and used to make phosphorus fertilizer because it is always found in chemical combination. Phosphorus is the most common word used to describe the element in agricultural fertilizer. However, in chemical terminology, it is proper to say that phosphoric acid (H_3PO_4) not elemental phosphorus is the chemical that changed agriculture.

Phosphorus exists in inorganic and organic forms in soil. The primary inorganic forms are $H_2PO_4^-$ and HPO_4^{2-}, which are available to and used by plants. Both can become unavailable when absorbed to soil colloids. Fifty to 80% of soil phosphorus is organic, which is derived from microbial breakdown of plant material incorporated into soil. Soil phosphorus is naturally available to plants, but the usual quantity is insufficient to support good growth or optimum crop yields. It can be lost from soil through wind or water erosion. Therefore, to achieve desired crop yields, phosphorus fertilization is required. Further description of phosphorus cycling in soil can be found in

Fate of Phosphorus Fertilizers in Soils, published by the New Zealand Institute of Chemistry.[1]

Phosphorus is a component of the complex nucleic acid structure of plants, which regulates protein synthesis. It is integral to cell division, cell membrane structure, development of new tissue, photosynthesis, and hormone activity in plants and animals, and is essential for all of any plant's complex energy transformations. Plants deficient in phosphorus are stunted, often have an abnormal dark-green color, or if sugars accumulate, anthocyanin[2] pigments can develop and produce reddish-purple colored foliage, which is normal in some plants. These severe deficiency symptoms are usually only observed in extremely low-phosphorus soils. Adding phosphorus fertilizer to soil low in available phosphorus promotes root growth, winter hardiness, stimulates tillering,[3] and often hastens maturity.

Yellow, unthrifty plants may be phosphorus deficient because of cold temperatures, which affect root extension and phosphorus uptake from soil. Delayed maturity of wheat is a typical deficiency symptom. Phosphorus deficiency causes stunted, sick-looking plants that produce lower quality fruit or flowers.

If adding phosphorus fertilizer to soil low in phosphorus is a good thing to do to prevent deficiency symptoms and promote plant growth, one must define low phosphorus to know if phosphorus fertilizer is needed. The answer is complicated by soil pH. Phosphorus availability is lower in strongly acidic (pH $\leq$5) and strongly alkaline (pH $\geq$8) soils because of increased reactivity of phosphorus and formation of insoluble compounds with aluminum and iron in acid soils and with calcium in alkaline soils. The pH associated with maximum phosphorus availability in soil usually is between 6.0 and 7.0. The accepted concentration (parts per million (ppm)) of low, optimum, and high phosphorus is dependent upon the crop to be grown. However, generally accepted concentrations using the Olsen et al. (1954) test are: very low, $\leq$5 ppm (mg/kg); low, 5–10 ppm; marginal, 10–17 ppm; optimal, 17–25, and high, $\geq$25 ppm. Soil retention and thus concentration are also affected by clay content and type of clay (the clay minerals—smectites (e.g., montmorillonite)) and kaolins (e.g., kaolinite).

A wide range of chemicals can be made by subtracting one or more protons from phosphoric acid (Figure 5.1). For many years, phosphorus has been one of the top 10 industrially processed chemical products, most of which is used to make fertilizer, as it has been for decades. When phosphoric acid is incorporated into biological systems, it is usually present as a phosphate ion (Öhrström, 2013). Without phosphoric acid, which is readily soluble in water, living systems would

---

1. http://www.nzic.org.nz/ChemProcesses/soils/index.html and http://scifun.chem.wisc.edu/chemweek/pdf/phosphoric_acid.pdf. Accessed January 2014.
2. Anthocyanins are water-soluble pigments that give flowers and other plant parts colors ranging from blue to shades of red.
3. Tillers are plant shoots that originate from roots or the bottom of the original stalk.

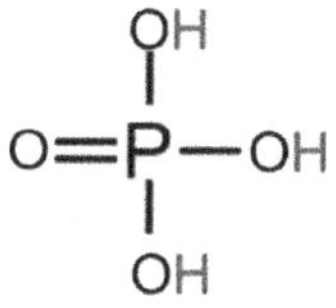

$$O=\overset{\overset{\displaystyle OH}{|}}{\underset{\underset{\displaystyle OH}{|}}{P}}-OH$$

$$H_3PO_4$$

FIGURE 5.1 Structure of orthophosphoric acid.

TABLE 5.1 Elemental Composition of the Human Body

Chemical	Percent	Atomic
Element of Mass Percent		
Oxygen	65	24
Carbon	18.5	12
Hydrogen	9.5	62
Nitrogen	3.2	1.8
Calcium	1.5	0.22
Phosphorus	1.1	0.78

be impossible. The principal component of mammalian bones and teeth is calcium phosphate—$Ca_{10}(PO_4)OH$. It is the second most abundant element (after calcium) present in our bodies. Calcium is 40% by weight of bones and teeth and phosphate ions are almost 60%. Whether calculated as percent of mass or atomic percent, about 1% of an adult's total body weight is phosphorus (Table 5.1).

Phosphorus is present in every cell, but 85% is found in bones and teeth. The food crops we depend upon and all other living systems have DNA. The acid in DNA is phosphoric acid. Thus, it is an essential component of life's structure, nutrition, and an integral part of the genome of all living systems.

The main food source of phosphorus is protein (meat and milk). Whole-grain breads and cereals contain more phosphorus than those made from refined flour. Its storage form (phytin[4]) is not digestible by humans. Fruits and vegetables contain only small amounts of phosphorus.[5] We obtain phosphorus through our food, much of which has been fertilized with mineral or organic fertilizers. But the long-term supply, sustainability, and the possibility of finding new sources of phosphorus required to fertilize and grow our food are not, but should be,

---

4. Phytin is the calcium or magnesium salt of phytic acid. Phytic acid (inositol hexakisphosphate) is a saturated cyclic acid. It is the principal phosphorus storage molecule in many plant tissues.
5. http://www.nlm.nih.gov/medlineplus/ency/article/002424.htm. Accessed November 2013.

topics of debate and intense investigation. We simply assume that because it has always been available, it always will be.

HISTORY

Justus F. von Liebig (1803–1873), a German chemist, made major contributions to agricultural and biological chemistry. He is credited with originating laboratory-oriented teaching of chemistry and for discovering that nitrogen is an essential plant nutrient. Liebig is commonly, but erroneously, regarded as the formulator of the Law of the Minimum, which described the effect of individual nutrients on crop yield. Liebig's Law of the Minimum states that plant growth and crop yield are a function of and proportional to the amount of the essential nutrient present in the lowest concentration. Sprengel (1787–1859), a German botanist, was actually the first to formulate the theory of the minimum in agricultural chemistry, which was popularized as a scientific concept by Liebig. Liebig's (Sprengel's) law is commonly depicted as a barrel (Figure 5.2). Liebig wrote, "The crops on the field diminish or increase in exact proportion to the diminution or increase of the mineral substances conveyed to it in manure." The intent of the barrel is to show that the nutrient present in the lowest amount (the lowest barrel stave) required for full growth will be the primary nutritional limit to plant growth. Liebig contributed greatly to our understanding of plant nutrition. He was the first to denounce the vitalist theory of humus: "Living organisms are fundamentally different from non-living entities because they contain some non-physical element or are governed by different principles than are inanimate things.[6]" He claimed that ammonia (nitrogen) was the most important nutrient and later promoted the importance of inorganic minerals. In England,[7] he attempted to verify his theories with a fertilizer he created by treating phosphate of lime in bone meal with sulfuric acid. It was less expensive than other available sources, but failed because the phosphorus was not absorbed by crops. His work showed that phosphate of lime in bone meal could be rendered more readily available to plants by treatment with sulfuric acid.

Sir John Bennet Lawes (1814–1900) was an English entrepreneur and agricultural scientist[8] who surely knew of Liebig's work. Lawes founded an experimental farm at his home, Rothamsted Manor, in the English county of Hertfordshire that eventually became the Rothamsted Experimental Station. The experiment station still exists and includes the Park Grass Experiment, the world's oldest and longest running agricultural field experiment. It is not, however, the world's oldest, continuing scientific experiment. That distinction

---

6. http://en.wikipedia.org/wiki/Vitalism. Accessed August 2014.
7. http://en.wikipedia.org/wiki/History_of_fertilizer. Accessed November 2013.
8. http://en.wikipedia.org/wiki/John_Bennet_Lawes. Accessed November 2013.

FIGURE 5.2   Liebig's barrel (http://en.wikipedia.org/wiki/Liebig's_law_of_the_minimum).

belongs to the Oxford electric bell at Oxford University, which has been ringing continuously since it was first displayed in 1840.[9]

The Park Grass experiment, established in 1856, has been observed and monitored ever since. It was established to study the effect of different fertilizers on soil fertility, forage yield, and the biological composition/biodiversity of the original and developing vegetation. About 1837, Lawes began to study the effects of various manures (any natural or artificial substance used to fertilize soil) on plants growing in pots, and a year or two later he extended the study to crops in the field. In 1842, he patented a manure formed by treating phosphate rock with sulfuric acid (similar to Liebig's earlier work), and thus initiated the artificial manure industry (the agricultural fertilizer industry). He was the first to manufacture superphosphate at his factory in Deptford, England, in 1842. Lawes and J. H. Gilbert worked together on experiments on raising crops and feeding animals for more than half a century. Their work and subsequent publications have made the Park Grass Experiment and Rothamsted famous throughout the agricultural scientific community.

SOURCES

There are no alternatives to natural phosphorus. It is an element and cannot be synthesized (created) in a laboratory. Phosphorus is more abundant than nitrogen in the earth's crust, but the amount on earth is finite, although new, undiscovered sources can be found, and it can be recycled. It is mined primarily as common phosphate rock, which is found worldwide. A few countries (see Table 5.2),

---

9. http://www.improbable.com/airchives/paperair/volume7/v7i3/long-run-7-3.html. Accessed November 2014.

TABLE 5.2 World Phosphate Reserves

Country	Million MT of Phosphate Rock Reserves
Morocco	5700
China	3700
South Africa	1500
Jordan	1500
United States	1100
Brazil	260
Russia	200
Israel	180
Syria	100
Tunisia	100
All other countries	1660
Total	16,000

control nearly 60% of the world's known phosphorus supply. After mining, the rock is processed, formulated into fertilizer, and applied to soil. Phosphorus does not cycle between plants/and the atmosphere. It does cycle between plants and soils and is lost through soil erosion and water runoff into creeks, rivers, and ultimately the ocean (Batjes, 2011).

The Ashgate Handbook of pesticides and agricultural chemicals (Milne, 2000) lists 55 different proprietary brands (sources) of phosphorus fertilizer. Most include nitrogen and potassium, which make them more complete fertilizers. A few are only phosphorus fertilizers: superphosphate ($CaH_4O_8P_2$) contains 25–28% soluble phosphate[10]; Tennessee phosphates contain 60–70% of the relatively insoluble calcium phosphate—$Ca_3(PO_4)_2$ (the primary natural form; water solubility = 0.002 g/100 g); triple super phosphate contains 48–49% P_2O_5, which provides 41–42% water-soluble P_2O_5; and African phosphates, found in Tunisia and Algeria, contain 55–65% calcium phosphate. Superphosphate is produced by the action of concentrated sulfuric acid on powdered phosphate rock:

$$Ca_3(PO_4)_{2(s)} + 2H_2SO_{4(aq)} = 2CaSO_{4(aq)} + Ca(H_2PO_4)_{2(aq)}$$

---

10. The units used to express phosphorus content can be confusing. Farmers use phosphorus fertilizer. Other terms include elemental P; the phosphate anion, PO_4; and phosphorus oxide, P_2O_5. Fertilizer's content of elemental phosphorus P can be determined by $P = P_2O_5/2.29$; the content of phosphorus oxide can be determined from $P_2O_5 = 2.29 \times P$.

Triple superphosphate (originally known as double superphosphate) is produced by the action of concentrated phosphoric acid on ground phosphate rock:

$$Ca_3(PO_4)_{2(s)} + 4\ H_3PO_{4(aq)} = 3\ Ca^{2+}_{(aq)} + 6\ H_2PO_4^{1-}_{(aq)} = 3\ Ca(H_2PO_4)_{2(aq)}$$

The active ingredient, monocalcium phosphate, is identical to that of superphosphate, but does not included calcium sulfate that is formed when sulfuric acid is used. The phosphorus content of triple superphosphate (17–23% P; 44–52% P_2O_5) is roughly three times greater than superphosphate (7–9.5% P; 16–22% P_2O_5). Triple superphosphate was the most common phosphate (P) fertilizer in the United States until the 1960s, when ammonium and di-ammonium phosphates—$(NH_4)_3PO_4$—became popular because of the added nitrogen.[11] Total world phosphate fertilizer production is 38–41 million metric tons (MT) P_2O_5/year and 80% is based on phosphoric and sulfuric acid production (Gregory et al. (2010)).

Prior to the twentieth century, farmers relied almost exclusively on natural levels of soil phosphorus, which was often, perhaps routinely, supplemented by application of animal manure. Carolan (2011) reports that England imported bones from Europe in the early nineteenth century because they were a source of phosphorus. In the 1840s, scientists found that coprolites (stony, petrified fecal matter of animals) could be dissolved in sulfuric acid to produce what became known as superphosphate.

Baron Alexander von Humboldt (1769–1859), a Prussian geographer, naturalist, and explorer, described Latin America between 1799 and 1804 from a scientific point of view. His descriptions and findings were published in a number of volumes over 21 years, which, in a real sense, became the foundation of the sciences of physical geography and meteorology. Using his 1817 delineation of the world's isothermal lines, he devised the means of comparing the climatic conditions of various countries.[12] Humboldt was the first European to recognize the importance of preserving the quinine tree (*Chinchona officinalis* L.) because its bark was the only source of the treatment for malaria (it did not cure malaria). The active ingredient, quinine, was discovered later (see Chapter 7). He encountered guano (a Quechua word) in 1802, in deposits up to 150 feet deep, on the Chincha islands off the coast of Peru. After investigating its fertilizing properties at Callao in Peru, he was the first to introduce guano to Europe. His subsequent writing made it well known in Europe. During the guano boom of the nineteenth century, the vast majority of seabird guano was harvested from the Peruvian guano islands. Large quantities were also exported from the Caribbean, atolls in the Central Pacific, islands off the coast of Namibia, Oman, Patagonia, and Baja California. At that time, massive deposits (as much as 50 feet deep) of guano

---

11. http://en.wikipedia.org/wiki/Monocalcium_phosphate#Superphosphate. Accessed November 2013.

12. http://en.wikipedia.org/wiki/Alexander_von_Humboldt. Accessed November 2013.

existed on some islands. The American Guano Company was formed in New York City in September 1855 to collect and import the fertilizer. Guano was selling for $55/ton. The company expected profits of $2.4 million/year.

As mentioned in Chapter 4, Peru and Bolivia both had large reserves of high-quality nitrate and phosphorus rich deposits of guano. In the late 1800s, Bolivia's boundaries extended to the Pacific coast. Guano's value as fertilizer made the Atacama Desert (in Northern Chile) strategically and economically important. Bolivia, Chile, and Peru each claimed the area, which had the world's largest reserves of guano and thus of fertilizer. The selected solution was the Guano War of 1879–1883, also known as the Second War of the Pacific. Bolivia lost its Pacific coastal area and its guano reserves. Subsequently, guano was called Peruvian or Chilean guano because Bolivia lost the war.

Guano typically contains 8–16% nitrogen (the majority of which is uric acid), 8–12% equivalent phosphoric acid, and 2–3% equivalent potash. It is the excrement of seabirds, cave-dwelling bats, and pinnipeds (a suborder of carnivorous aquatic mammals (e.g., seals, walruses)). It was regarded as so important that the US Congress passed the Guano Islands Act in 1856, which gave US citizens who discovered a source of guano on an unclaimed island exclusive rights to the deposits. More than 100 islands were claimed as US possessions under the Act, but it did not specify the islands fate after private interests took all the guano. At least 12 islands, now devoid of guano, are still US territories.[13]

In the nineteenth century, guano trade played a pivotal role in the development of modern farming practices. Two factors quickly diminished and eventually eliminated guano imports. The first is similar to what happens to many natural resources present in finite quantities, it was overexploited. The second, more important development was the replacement of guano with manufactured phosphate fertilizer derived from phosphate rock.

SUPPLY

The main source of phosphorus is sedimentary rock containing the calcium phosphate mineral apatite (a group of phosphate minerals). Phosphate rock is mined and either solubilized to produce phosphoric acid or smelted to produce elemental phosphorus. Phosphoric rock is treated with phosphoric acid to produce triple superphosphate or with anhydrous ammonia to produce ammonium phosphate fertilizers. Approximately 90% of phosphate rock production is used for fertilizer and animal feed supplements. The remaining 10% has several minor, but nevertheless important uses.[14] Phosphate products are mass-produced and available cheaply in large quantities. In addition to fertilizer they are used in many ways:

---

13. For further information, see: http://www.ushistoryscene.com/uncategorized/guanoislandsbirdturd and http://en.wikipedia.org/wiki/Guano_Islands_Act. Accessed August 2014.

14. http://en.wikipedia.org/wiki/Phosphoric_acid. Accessed December 2013.

- Rust removal by direct application to metals.
- Food-grade phosphoric acid is used to acidify foods and colas to provide a tangy/sour taste.
- Phosphoric acid is used in dentistry as an etching solution to clean and roughen the surfaces of teeth where dental appliances or fillings will be placed.
- The acid is also used in many teeth whiteners to eliminate plaque before application.
- It is also a minor ingredient in over-the-counter antinausea medication.

Each of these uses has provoked controversy about possible detrimental effects on human health.

Modern agricultural systems are dependent on continual inputs of phosphorus fertilizers processed from phosphate rock, the world's primary source. Phosphate rock (phosphorite or rock phosphate) contains 15–20% phosphate bearing minerals. The phosphorus is present as fluorapatite—$Ca_5(PO_4)_3F$—or as hydroxyapatite (apatite)—$Ca_5(PO_4)_3OH$. Another source is dissolved phosphate minerals from igneous and metamorphic rocks. Phosphate rock is a nonrenewable resource that takes 10–15 million years to cycle naturally. All crops need phosphorus, and most must be supplied in fertilizer because natural soil supplies are too low to fill crop demand. Five countries control around 90% of the world's remaining phosphate rock reserves: China, Jordan, Morocco (which controls the Western Sahara and nearly 40% of the world's known reserves), South Africa, and the United States. Morocco has a greater percentage (35%) of world reserves, but China (23%) now produces more phosphorus fertilizer (approximately 57 MT/year[15]) than the United States (30 MT), Morocco (20 MT), Russia (5 MT), and all other producers (44 MT). The projected world reserves are shown in Table 5.2.

The Morocco-based phosphate producer—Office Chérifien des Phosphates (OCP)—controls nearly all of Morocco's phosphate production. The OCP reports that Morocco is strengthening its production capacity. In accordance with the OCP's development strategy, its production capacity of phosphate rock will be 50 million MT and phosphate fertilizer could reach 9 million MT in 2020. Since 1995, Morocco has exported more than 10 million MT/year of phosphorus ore (50% of global exports) and 1.87 million MT of diammonium phosphate in 2010, accounting for 11.9% of global exports.[16] Morocco could satisfy demand by increasing production. It would thereby obtain a much greater share of worldwide production, from 15% in 2010 to potentially as much as 80% by 2100, and, inevitably, more control over market prices (Cooper et al., 2011). If this happens, Morocco's share of global reserves will increase, from 77% in 2011 to 89% by 2100.

The US Geological Survey says that of the 65 billion tons of the world's known phosphate rock reserves—and the estimated 16 billion tons that might be

15. MT = metric ton = 1000 kg/2205 pounds (usually rounded to 2200 pounds).
16. http://www.kcomber.com/publishing/home/2013/02/01/33/morocco-s-production-capacity-in-phosphate-products-is-expected-to-rise.html. Accessed August 2014.

cost effective to mine—almost 80% is in the Western Sahara and Morocco. Add in China's reserves, and the figure rises to almost 90%.

The United States produces about 30 million MT of phosphate rock each year, which should last 40 years, assuming today's rate of production (Vaccari, 2009). Cooper et al. (2011) conclude that, unless additional sources of phosphorus are discovered or phosphorus recycling increases significantly, future global phosphorus security may be increasingly reliant on a single country. Since 2010, deposits of at least 1 gigaton (1 billion tons) have been discovered in Morocco, the Western Sahara, Algeria, and Iraq, but there is a dearth of data on their ease of harvest or phosphorus content.

For predicting future supply, the ultimately recoverable resource is commonly used. It is equal to the combined sum of all historic and predicted future production. Estimates of the ultimately recoverable resource of phosphorus are highly variable (Table 5.3) (Ward, 2008), which highlights the uncertainty about the amount of phosphorus-bearing material available. Future production projections also have a wide variation (Mohr and Evans, 2013). External drivers (drought, famine, human population, war) will inevitably influence annual production and thus the lifetime of recoverable resources. Clearly, there is considerable uncertainty in the supply and potential future production of this crucial agricultural, indeed societal resource, which is a nice way of saying, No one knows![17]

Known reserves might last 300–400 years at current world production rates of 160–170 million MT of fertilizer/year. Because phosphorus fertilizer production is expected to increase by 2–3%/year and use by as much as 5% during the next 5 years, the life expectancy of reserves could be much shorter (Heffer and Prud'homme 2010).

In spite of the wide range of predictions of phosphate rock supply, others have attempted to predict when production will peak or if it already has peaked (Table 5.4). Peak production refers to the time at which maximum global production is reached, after which it will only decline. This concept is applicable to many finite natural resources extracted by humans. It has been regularly applied to oil in recent years. Proponents of the theory argue that, because the resource is finite and demand is increasing, the supply must decrease with time. The important question is not when the world will run out of phosphate reserves completely, but when it will run out of high grade, easily accessible phosphate rock (Schroeder et al., 2010) and how that will affect the price of this essential agricultural resource. To date, nearly 80% of mined phosphate rock has been high quality, sedimentary rock, but the quality and quantity of available reserves is declining. The cost of extracting and processing phosphorus from igneous rock is higher (Gregory et al., 2010). Therefore, it is logical to assume that the cost of producing phosphoric acid may increase by as much as 30–40% (Gregory et al., 2010). The market

---

17. http://www.philica.com/display_article.php?article_id=380. Accessed December 2013.

TABLE 5.3 Estimates of Global Phosphate Reserves and Years to Depletion

Source	Year of Estimate	Reserves Estimated Million Metric Tons	Years to Depletion
Déry and Anderson	2007	1000	
Cordell	2009	3200	
USGS: Jasinski	2010	16,000	
IFDC:-Van Kauwenbergh	2010	60,000	300–400
Van Kauwenbergh	2010[b]	290,000–490,000[a]	
Heffer and Prud'homme	2010	≤300–400	
Van Vuuren	2010	1600–36,500	
Global phosphorus res. Initiative (GPRI)	2013[c]	75–200	
Cooper et al.	2011	100	
USGS	2011	17,630–71,650	<80
USGS: Jasinski	2012	71,000	
Sanchez	2013	300–400	

[a]An estimate of known and potential reserves.
[b]http://www.unep.org/yearbook/2011/pdfs/phosphorus_and_food_productioin.pdf. Accessed December 2013.
[c]http://phosphorusfutures.net/. Accessed January 2014.

adjustment will be difficult for producers and users of phosphorus fertilizer. It will favor additional production mainly from existing large sedimentary rock processors with access to lower quality rock. As with nitrogen fertilizer, these cost increases will provide incentive to lower fertilizer processing costs and improve field-use efficiency. Because phosphorus is one of the three major elements (plus nitrogen and potassium) critical to plant growth, dire consequences for world agricultural production and food security are linked to peak phosphate (Van Kauwenbergh, 2010).

Approximately 161 MT of phosphate rock were mined in 2011, and 70% of global production is currently produced from reserves (Cooper et al., 2011). More than 50% is used in Asia where 45% is produced. Increasing world demand for agricultural fertilizer will soon result in a significant global production deficit, which by 2070 may be larger than production.

Thus, one must, at a minimum, consider the validity and accuracy of the conservative estimate of world's reserves and possible peak production. Although

TABLE 5.4 Estimates of Peak Phosphorus Production

Source	Year	Peak Year(s)	Peak Production, Million MT/Year
Déry and Anderson	2007	1988	20
Cordell, Dranger, and White	2009	2034	29
Jasinski	2010, 2012	1987–1988/2008	
Van Vuuren, Bouwman, and Beusen	2010	1988–2100	66–115
Van Kauwenbergh	2010	2033–2034	
Cooper et al.	2011	Within 100	

http://en.wikipedia.org/wiki/Peak_phosphorus. Accessed December 2013.

there will undoubtedly be more phosphate discoveries, it is likely, just as it is with oil, that they will require progressively more effort and expense to extract and refine, and production will be on an inexorable path of decline.[18]

Not all agree with estimates of reserves or peak production time. Sanchez (2013), director of the Agriculture and Food Security Center of Columbia University's Earth Institute, does not believe there is a shortage of phosphorus. "Once every decade, people say we are going to run out of phosphorus. Each time this is disproven. All the most reliable estimates show that we have enough phosphate rock resources to last between 300 and 400 more years." Sanchez claims that new research shows that the amount of phosphorus coming to the surface by tectonic uplift is in the same range as the amounts of phosphate rock we are now extracting. Sanchez's view is consistent with the estimates of Van Kauwenbergh (2010) and Heffer and Prud'homme (2010), shown in Table 5.3.

If price is an indicator of supply, it is worthy of note that in mid-2008 the price of phosphate rock was $430.00/MT, 800% higher than in early 2007. However, a snapshot is usually not an accurate measure. In 2000, the price was $48/MT, it peaked in 2008, dropped to $95 in 2010, and in July 2013 it was $121/MT.

Debates about phosphorus supplies are similar to those about oil, the world's primary energy source. The major difference is that some suggest we may never run out of oil (Mann, 2013) and substitutes are being developed (solar, wind, biomass). But there is no substitute for phosphorus in living systems and in food production. Phosphorus cannot be manufactured, though fortunately it can be

---

18. http://web.mit.edu/12.000/www/m2016/finalwebsite/problems/phosphorus.html. Accessed November 2013.

recovered and reused over and over again. There are well-known ways. If we got serious about recycling our bio-waste, we could reduce our need for phosphate rock. It turns out that urine and feces contain a lot of phosphorus—which is why they make good fertilizer.

AGRICULTURAL ROLE

As the Earth's finite supply of phosphorus diminishes, farmers—that is the entire agricultural system—will be less able to accomplish their primary goal: producing food; people will starve. As mentioned, there is great uncertainty about when that will happen. It depends on the rate of use, discovery of new sources, the potential of recovery and reuse, and the potential increased cost of processing new sources. Because all types of agriculture are based on living systems, none can survive without phosphorus, nearly all of which is supplied by manufactured fertilizer, which uses ±90% of the world's phosphorus (Christen, 2007, p. 98). It is essential to modern agriculture and to life. It takes about 1 ton of phosphate to produce 130 tons of grain, although this is highly variable and dependent on soil, farm history, crop type, and fertilizing efficiency (Vaccari, 2009). No one really knows how severe the effect of severely diminished phosphorus supply may be on production agriculture, but all agree it will be catastrophic. Facing phosphorus scarcity is the sole subject of the 2014 Virtual Issue 5 of Plant and Soil, the international journal of plant–soil relationships.

Batjes (2011), with significant supporting scientific evidence (Buresh et al., 1997; Stoorvogel et al., 1993), cites soil fertility degradation as the single most important threat to food security in sub-Saharan Africa, a part of the world where hunger dominates the lives of many people. There food production is low, in soils with low available phosphorus, and production remains low because phosphorus output in harvested crops is not replaced by fertilizer input. Farmers cannot afford the fertilizer, and because the farming areas are often remote, transport and distribution of fertilizer costs more. In contrast, some of the most fertile soils in the world are overfertilized because farmers can afford it and seek yield certainty. Continued application of more phosphorus than crops use increases soil fertility, but the adsorptive capacity of soil and soil management practices can lead to two undesirable consequences. A significant portion of the added phosphorus will not be available to crops. Second, excess phosphorus may exceed a soil's capacity to retain it and thereby lead to runoff, eutrophication, and water pollution.

In 1860, 32,000 tons of synthetic fertilizer were used in the United States. It quickly replaced guano. By 1972, 41 million tons of complete (N, P, K) fertilizer were applied in the United States (Martin et al., 1976, p. 151). Following World War II, fertilizer use expanded rapidly in the United States, but leveled off in the early 1980s after reaching a peak of 23.7 million nutrient tons in 1981.[19] Corn production accounts for half of US fertilizer use. One-third of the US corn crop

---

19. A nutrient ton is calculated on the basis of the N, P_2O_5, and K_2O content. It does not include nonnutrient components of fertilizer.

(approximately 4.7 billion bushels) is used to produce ethanol, nearly all of the rest is fed to livestock. The ethanol is not used by the distilling industry to make alcoholic drinks; it is added to gasoline for your car. Why is ethanol in your gasoline? In the early 1990s, the United States amended the Clean Air Act of 1970 to require use of oxygenated gasoline because the fuel burned more completely. The first chemical used was methyl tert-butyl ether, or MTBE. It was cheap and achieved higher octane ratings. California scientists discovered in 1995 that MTBE was showing up in high concentrations in drinking water and traced it back to spilled gasoline and leaky underground containers. Ethanol was seen as a safer replacement. Its use was pushed by the agricultural industry and the requirement for fertilizer (primarily nitrogen and phosphorus) increased because more corn was grown.

In 1960, 1.06 million tons of all forms of superphosphate were used in United States. Use peaked in 1970 at 1.7 million tons and has steadily declined since then to 347,298 tons in 2010. At the same time the use of di-ammonium phosphate (N + P) has increased steadily from 20,000 tons in 1962 to a peak of 3.5 million tons in 1990 and 2.3 million tons in 2010.[20] The decline was due to farmers' recognition of maximum yield from reduced rates and environmental concern about runoff and subsequent pollution and eutrophication of water. Worldwide, 18.1 million tons of P_2O_5 were used in 2011 (Gregory et al., 2010; IFA, 2009).

The more important estimate is that as much as 40–60% of world food production is dependent on synthetic fertilizers. When anything is essential to some activity, debating its importance relative to some other essential ingredient seems rather silly. If the amount used is the primary criterion of importance, then nitrogen is the most important fertilizer ingredient. However, nitrogen is not a limited resource; phosphorus is. Global demand for phosphorus is projected to increase between 50% and 100% by 2050 because of increased food production, increased meat consumption as societies become richer, and the presently unpredictable growth in biofuel production (Cordell et al., 2009, p. 293). At current production levels, the world will exhaust the known, easily obtained and processed reserves; Tables 5.2 and 5.3 indicate we really don't know when. Several scientists agree that without new sources for high quality easily mined phosphorus, agriculture will face major production problems within the next 50–100 years.

If those who practice and serve agriculture are to meet their primary production and moral obligation, they must be aware of and deal with an increasing demand and what many believe to be the inexorably decreasing supply of phosphorus. As mentioned, it is likely, although not certain, that more phosphorus reserves will be discovered. It is also likely, but not certain, that new supplies will require more effort and expense to produce phosphorus fertilizer.[21]

---

20. http://www.ers.usda.gov/data-products/fertilizer-use-and-price.aspx#26720. Accessed November 2013. Some data are presented in metric tons (2205 pounds) and other in tons (2000 pounds), commonly called short tons.
21. http://web.mit.edu/12.000/www/m2016/finalwebsite/problems/phosphorus.html. Accessed November 2013.

However, the world's consumption level—the demand for food—will continue to rise, creating an ever-widening supply–demand gap. That is how phosphorus has and is likely to transform agricultural productivity and agricultural practice—how we farm.

REFERENCES

Batjes, N.H., 2011. Global Distribution of Soil Phosphorus Retention Potential. International Soil Reference and Information Centre (ISRIC) – World Soil Information (with dataset), Wageningen. ISRIC Report 2011/06, 42 pp.

Carolan, M., 2011. The Real Cost of Cheap Food. Earthscan, London, UK, New York. 272 pp.

Christen, K., 2007. Closing the phosphorus loop. Environmental Science and Technology 41 (7), 2078.

Cooper, J., Lombardi, R., Boardman, D., Carliell-Marquet, C., 2011. The future distribution and production of global phosphate rock reserves. Resources Conservation and Recycling 57, 78–86.

Cordell, D., Dranger, J.-O., White, S., 2009. The story of phosphorus: global food security and food for thought. Global Environmental Change 19 (2), 292–305.

Déry, P., Anderson, B., August 13, 2007. Peak Phosphorus. Energy Bulletin.

Gregory, D.I., Haefele, S.M., Buresh, R.J., Singh, U., 2010. Fertilizer use, markets, and management. In: Pandey, S. (Ed.), Rice in the Global Economy: Strategic Research and Policy Issues for Food Security. International Rice Research Institute, Los Baños, Philippines, pp. 231–263.

Heffer, P., Prud'homme, M., 2010. Fertilizer Outlook 2010–2014. In: 78th IFA Annual Conference, May 31–June 2, International Fertilizer Industry Association (IFA), Paris.

IFA (International Fertilizer Industry Association), 2009. Assessment of Fertilizer Use by Crop at the Global Level: 2006/07–2007/08. International Fertilizer Industry Association. 11 pp.

Jasinski, S.M., January 2010. Minerals Commodities Summary: Phosphate Rock. http://minerals.usgs.gov/minerals/pubs/commodity/phosphate_rock/mcs-2010-phosp.pdf (accessed November 2013).

Jasinski, S.M., January 2012. Minerals Commodities Summary: Phosphate Rock. Retrieved from http://minerals.usgs.gov/minerals/pubs/commodity/phosphate_rock/mcs-2012-phosp.pdf (accessed November 2013).

Mann, C.C., 2013. What if We Never Run Out of Oil? The Atlantic. May p. 48, 50–54, 56, 58, 60–61.

Martin, J.H., Leonard, W.H., Stamp, D.L., 1976. Principles of Field Crop Production, third ed. McMillan Publishing Company, Inc, New York. 1118 pp.

Milne, G.W.A., 2000. Ashgate Handbook of Pesticides and Agricultural Chemicals. Ashgate Publications, Ltd, Hampshire, England. 206 pp.

Mohr, S., Evans, G.M., 2013. Projections of Future Phosphorus Production. Article 380. 47pp. Also see: Chemo-Philica.comhttp://www.philica.com/display_article.php?article_id=380 (accessed November 2013).

Öhrström, L., June 19, 2013. Phosphoric Acid. Chemistry World. Scifun.chem.wisc.edu/chemweek/chemweek.html.

Olsen, S.R., Cole, C.V., Watanabe, F.S., Dean, L.A., 1954. Estimation of Available Phosphorus in Soils by Extraction with Sodium Bicarbonate. USDA circular 939. U.S. Govt. Printing Office, Washington, D.C.

Replenishing soil fertility and Africa. In: Buresh, R.J., Sanchez, P.A., Calhoun, F. (Eds.), 1997. Proceedings of an International Symposium Cosponsored by Divisions A-6 (International Agronomy) and S-4 (Soil Fertility and Plant Nutrition) and the International Center for Research in Agroforestry (Indianapolis, IN, November 6, 1996). American Society of Agronomy and the Soil Science Society of America, p. 251.

Sanchez, P., 2013. http://blogs.ei.columbia.edu/2013/04/01/phosphorus-essential-to-life-are-we-running-out/ (accessed December 2013).

Schroder, J.J., Cordell, D., Smit, A.L., Rosemarin, A., 2010. Sustainable use of phosphorous. Plant Research International. http://ec.europa.eu/environment/natres/pdf/sustainable_use_phosphorus.pdf (accessed November 2013).

Stoorvogel, J., Smaling, E.M.A., Janssen, B.H., 1993. Calculating soil nutrient balances in Africa at different scales. Fertilizer Research 35, 227–235.

Vaccari, D.A., 2009. Phosphorus: a looming crisis. Scientific American 300 (6), 54–59.

Van Kauwenbergh, S.J., 2010. World Phosphate Rock Reserves and Resources. Tech. Bul. IFDC-T-75. Int. Fertilizer Development Center, Muscle Shoals, AL, U.S. 60 pp.

Van Vuuren, D.P., Bouwman, A.F., Beusen, A.H.W., 2010. Phosphorus demand for the 1970–2100 period: a scenario analysis of resource depletion. Global Environmental Change 20 (3), 428–439.

Ward, J., August 26, 2008. Peak Phosphorus: Quoted Reserves vs Production History. Energy Bulletin.

Chapter 6

2,4-D: An Herbicide

What the atom bomb was to theoretical physics, the herbicide 2,4-dichlorophen-oxyacetic acid (2,4-D) was to weed control and mechanized farming.

Hilton (2007, p. 81).

Chapter Outline

A NEW CHEMICAL

In 1940, Robert Pokorny, a chemist, was employed to undertake exploratory chemical synthesis by the C. B. Dolge Company of Westport, CT, a specialty chemical company. It was a spinoff of the Embalmers Supply Company, which was managed by Arthur Dolge. The C. B. Dolge Company, formed in 1931, sold mostly arsenical herbicides. It was managed by Carl Dolge, Arthur's brother. The company also sold janitorial supplies such as soap and cleaning materials for institutions.[1] Eventually, the C. B. Dolge Company was sold to Rochester Midland company in upstate New York.

We will never know why Pokorny synthesized 2,4 dichlorophenoxy acetic acid (2,4-D, Figure 6.1) and 2,4,5-trichloroacetic acid (2,4,5-T, Figure 6.2).

---

1. Personal communication: Mr. R. S. Beck, August 2013.

Six Chemicals That Changed Agriculture. http://dx.doi.org/10.1016/B978-0-12-800561-3.00006-7

Chemical synthesis is one of the things organic chemists do. In the first case, equimolar quantities of 2,4-dichlorophenol (5.0 g) and monochloroacetic acid (2.9 g) were heated with a slight excess of sodium hydroxide (2.7 g) and 15 mL of water. In the second case 5.0 g of 2,4,5-trichlorophenol were used. The synthesis method used by Pokorny is essentially the same four-step method presently used to make these herbicides.[2] A pH of 10–12 and a temperature of about 105 °C are favorable reaction conditions (Kearney and Kaufman, 1975, p. 10).

1. Acidic phenol is neutralized with caustic soda.
   2,4-dichlorophenol ($C_6H_4Cl_2O$) + NaOH → sodium dichlorophenate (NaDCP) + H_2O + heat
2. Sodium monochloroacetate (NaMCA—$C_2H_2ClNaO_2$) produced in an exothermic reaction.
   Monochloroacetic acid (MCAA—$CH_2ClCOOH$) + NaOH → sodium monochloroacetate (NaMCA—$ClCH_2COONa$) + H_2O
3. NaDCP is condensed with NaMCA to produce Na 2,4-D, the sodium salt of 2,4-D.
   NaDCP + NaMCA → sodium 2,4-dichlorophenoxyacetate + NaCl
4. Na 2,4-D is aged by the slow addition of NaOH to ensure that the phenol content is less than 0.5 w/w%.

Sodium 2,4-D can then be esterified. An ester is an organic chemical in which the hydrogen of the carboxyl group is replaced with a hydrocarbon group

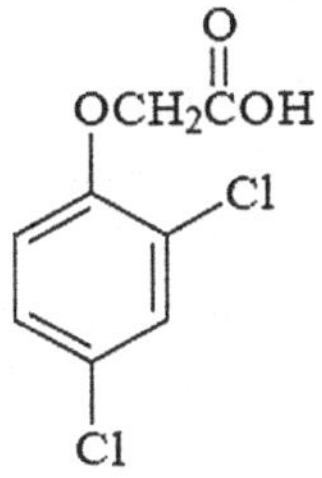

FIGURE 6.1 2,4-D: 2,4-dichlorophenoxyacetic acid.

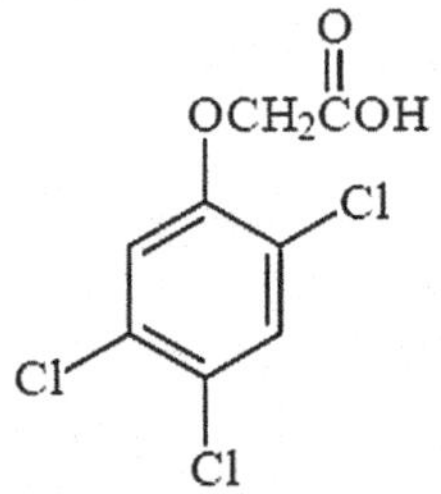

FIGURE 6.2 2,4-5-T: 2,4,5 trichlorophenoxyacetic acid.

---

2. http://nzic.org.nz/ChemProcesses/production/1J.pdf. Accessed January 2013.

(e.g., methyl [CH$_3$] or ethyl [CH$_2$CH$_3$]) or it may be converted to an amine salt (e.g., the dimethylamine salt [(CH$_3$)$_2$NH]).

2,4-D and 2,4,5-T were regarded as chemical curiosities and reported as new compounds in the *Journal of the American Chemical Society* (Pokorny, 1941). Only the synthesis was reported; there was no mention of any herbicidal activity. It may have been what the Dolge Company was looking for, but we will never know.

PLANT GROWTH REGULATORS

Charles Darwin, a name not usually associated with herbicides or weed control, reported in 1880 that plants always bend toward light. He believed, but did not prove, that leaf tips transmitted something to lower plant parts, which created the observed response. In the late 1920s and 1930s, extensive research on plant hormones laid the foundation for selective control of weeds and development of herbicides by the modern chemical industry. Plant physiologists followed Darwin's lead and began to look for an explanation for what came to be called phototropism. By 1929, Went and Thimann and others had determined that a light-sensitive growth hormone in seedling leaf tips controlled their growth. By 1926, the hormone had been extracted from plants (Kögl et al., 1934). Skoog and Thimann (1934) first observed that indoleacetic acid (IAA, Figure 6.3) promoted or inhibited plant growth depending on its concentration. In 1934, Went and Thimann synthesized the chemical and found that it affected plants exactly as the plant-produced hormone did.

Synthetic Auxins

There are at least six classes of hormones that affect plant growth: auxins, cytokinins, gibberellins, ethylene, abscisic acid, and polyamines. Plant hormones are chemicals that are produced in one location and act, in very low concentration, at another location. Auxins stimulate plant growth, particularly growth of excised coleoptile tissue. The name auxin generally refers to IAA; there are other active molecules. Gibberellins have varied effects on plant growth that differ between organs and between plants. They influence internode extension

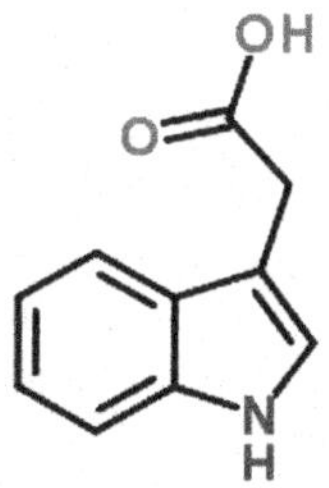

FIGURE 6.3   IAA: indole-3-acetic acid.

and thus can change dwarf to tall plants. They may affect cell division, induce fruit development, and can substitute for cold or light treatments required to induce sprouting or germination. There are no known herbicides whose primary mechanism of action is interference with gibberellin synthesis or action, but some carbamothioate herbicides may interfere with gibberellin biosynthesis as a secondary action.

Ethylene is a plant hormone involved in many aspects of growth. There are no herbicides whose primary mechanism of action is interference with ethylene action, although some nonherbicidal compounds have been developed to stimulate fruit ripening and stem growth of flowers. Auxin-like herbicides often increase ethylene production, which is linked to development of injury symptoms. There are no herbicides based on cytokinin, polyamine, or abscisic acid structure and no known herbicides that interfere with their action.

It is not possible to assign a specific physiological role to a compound within one of the five major hormone groups because they interact with each other and with other factors that influence plant growth. In a similar way, we do not know precisely how all herbicides mimic auxin action, but we know enough about them to use them intelligently. In most classifications, herbicides that interfere with plant growth include phenoxyacetic acids, benzoic acids, and picolinic acids. These growth regulator or hormone herbicides act at one or two specific auxin-binding proteins in the plasma membrane. They disrupt hormone balance, affect protein synthesis, and yield a range of growth abnormalities.

Growth Regulator Herbicides

Accounts vary about when the first work on growth regulator herbicides was done (Akamine, 1948). In 1935, Zimmerman and Hitchcock of the Boyce Thompson Institute (formerly in Yonkers, NY; now at Cornell University, Ithaca, NY) were investigating plant hormone-like substances. They particularly noted activity in phenyl and naphthyl acetic acids. They were the first to demonstrate that these molecules had physiological activity and affected cell elongation, morphogenesis, root development, and parthenocarpy (King, 1966).[3] In a 1942 paper, they described the substituted phenoxy acids (2,4-D is one) as growth regulators (auxin-like compounds) but did not report herbicidal activity. In 1938, V.C. Irvine, a chemist at the University of Colorado, reported that naphthoxyacetic acid was a very active plant growth regulator (Peterson, 1967). Zimmerman and Hitchcock (1935, 1942) also reported that the phenoxy acids were very active plant growth regulators. By 1939, they had discovered 54 different chemicals that affected plant growth when applied as vapors. "In April 1942, Zimmerman and Hitchcock reported the hormone-like responses (e.g., bending of stems and leaves (epinasty), swelling, induction of adventitious roots) that phenoxyacetic acids and benzoic acids induced when applied to plants in various forms"

---

3. Parthenocarpy: natural or artificially induced production of fruit without fertilization of ovules.

(Peterson, 1967). One of the most effective was 2,4-D, but they did not report on its herbicidal activity because they were studying chemicals that regulated plant growth. They were not looking for herbicides.

Similar work for similar reasons was done in Great Britain (Kirby, 1980). A more complete chronology and history of development of the hormone herbicides is available in Kirby (1980), Troyer (2001), and Zimdahl (2010). W.G. Templeman of Imperial Chemical Industries of Great Britain completed work in 1936 and 1937 that showed the toxic effects of naphthalene acetic acid ($C_{10}H_7CH_2CO_2H$) varied among species when whole plants were treated (Templeman, 1939; Troyer, 2001). In 1940, Troyer et al. showed that "growth substances applied appropriately would kill certain broad-leaved weeds in cereals without harming the crop". Troyer (2001) reported that Imperial Chemical Industries chemists had accomplished the synthesis of 2,4-D before its synthesis by Pokorny in 1941 and that Templeman and his colleagues had found 2,4-D and MCPA [(4-chloro-2-methylphenoxy)acetic acid] differed in their selectivity when used as herbicides in cereals. A British patent was filed in 1941, but the patent and scientific publications did not appear until after World War II ended, because the research was conducted under strict military secrecy (see the section World War II). Slade et al. (1945), in England, discovered that naphthaleneacetic acid at 25 lb/acre selectively removed charlock (wild mustard) from oats with little injury to oats. They also discovered the broad-leaved herbicidal properties of the sodium salt of MCPA = methoxone (King, 1966), a compound closely related to 2,4-D, and confirmed the selective activity of 2,4-D in publications that did not, because they could not, appear until after the war. When the war ended the chlorine-substituted phenoxyacetic acids [2,4-D, MCPA, and 2,4,5-T (2,4,5-trichlorophenoxyacetic acid)] were rapidly introduced as selective herbicides (Kearney and Kaufman, 1975). At this point, it is worthy of note that there are hundreds of chlorinated organic compounds that are used in many ways. Most, as pointed out by Le Couteur and Burreson (2004, p. 329), "are not poisonous, do not destroy the ozone layer, are not harmful to the environment, are not carcinogenic, and have never been used in warfare." Some, particularly among the pesticides, have been used inappropriately, perhaps stupidly is a better word. Some have been disposed of improperly (e.g., polychlorinated biphenyls) and have harmed the environment.

It is clear that work on the herbicidal properties of the phenoxyacetates in England preceded by several years the work in the United States. The British selected MCPA for further development not because of its herbicidal superiority, but because of the greater availability of the required chemical precursor chloro-cresol and the low availability of chloro-phenol in England (Norman et al., 1950). There was close involvement with Fanny Fern Davis of the US Golf Association greens section at Beltsville, MD. She knew of the work on selective control of dandelions in turf with 2,4-D and began extensive studies with 2,4-D. She was among the first to direct a program to develop practical, selective weed control with 2,4-D.

A Chicago carnation grower's question, "What is the effect of illuminating gas (acetylene) on carnations?" led to the eventual discovery of other plant growth regulating substances by Boyce Thompson scientists (King, 1966).

Auxin-like herbicides are effective because high tissue concentrations are maintained. They affect proteins in the plasma membrane, interfere with RNA production, and change the properties and integrity of the plasma membrane. The rate of protein synthesis and RNA concentration increase as persistent, auxin-like materials prevent normal and necessary fluctuation in auxin levels required for proper plant growth. Sugars and amino acids in reserve pools are mobilized by the action of auxin mimics. This is followed by, or occurs concurrently with, increased protein and RNA synthesis and degradation and depolymerization of cell walls. There are chemical structural requirements that must be satisfied for an herbicide to interfere with auxin activity. These include a negative charge on a carboxyl group, which must be in a particular orientation (spatial configuration) with respect to the phenoxy 6-carbon structure and have a partial positive charge associated with the ring that is at a variable distance from the negative charge. These spatial and charge requirements enable herbicide molecules to interact precisely with receptor proteins.

Use of these translocated, auxin-like herbicides offers significant advantages, but they have limitations. Advantages include the need for only small quantities and foliar application that can kill roots deep in soil because of phloem translocation. Low doses keep residual problems to a minimum. However, limitations are just as real and important. Only roots attached to living shoots in the right growth stage are killed. A uniform stage of growth is often required and very difficult to achieve with a variable plant population whose individuals emerge over time and grow at different rates. Residual effects can be important if soil remains dry after application.

Growth regulator herbicides are not metabolically stable in plants, are readily translocated, have low mammalian toxicity, are nonstaining, not flammable, do not persist long in the environment, and are metabolized to a variety of different chemical molecules. Growth regulators are not resistant to metabolism, but plants cannot control their concentration as they control concentration of natural plant hormones. This is an important reason for their activity. Physically, their action blocks the plant's vascular system because of excessive cell division and excessive growth with consequent crushing of the vascular transport system.

PHENOXY ACIDS

Structure

2,4-dichlorophenoxyaceticacid (Figure 6.1) is a white, crystalline solid that is slightly soluble in water. It, many other pesticides, and thousands of common chemicals including plastics, flavors (e.g., vanilla), marijuana, and explosives are based on phenol (Figure 6.4). Phenol is the generic name for a family of organic compounds characterized by a hydroxyl group attached to a carbon atom that is part of an aromatic ring. It is also the specific name for its simplest member, monohydroxybenzene (C_6H_5OH), also known as carbolic acid.

Soon after 2,4-D's herbicidal activity was discovered, it became obvious that substitution of hydrogen in the carboxyl group affected activity. Therefore, a great deal of research was done on formulation to develop the dominant ester, salt, and amine formulations (Figures 6.5–6.7, respectively).

These forms are important because of differences in their ability to penetrate plant cuticles and their volatility. In general, esters are more phytotoxic on an acid equivalent basis than are amine and salt forms. Technically, amines are salts, but are distinct because of their different chemical properties. Amines are, in general, soluble in water and used in aqueous concentrate formulations. Esters are oil-soluble, but may be applied as water emulsions with a suitable

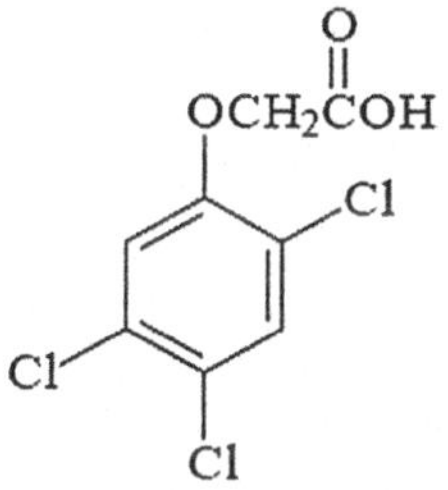

FIGURE 6.4 Phenol.

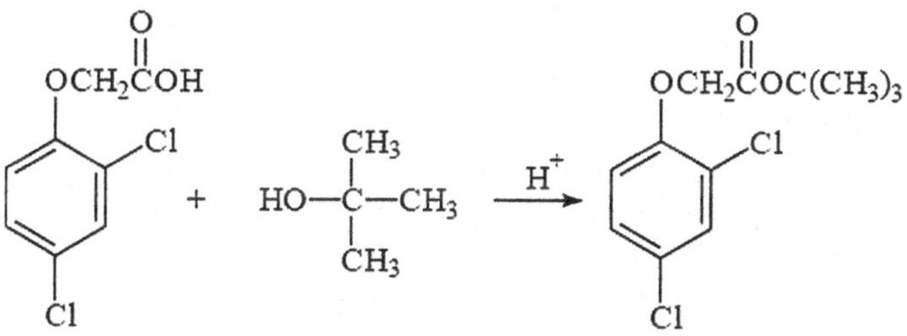

FIGURE 6.5 Tri-methyl ester of 2,4-dichlorophenoxyacetic acid (2,4-D).

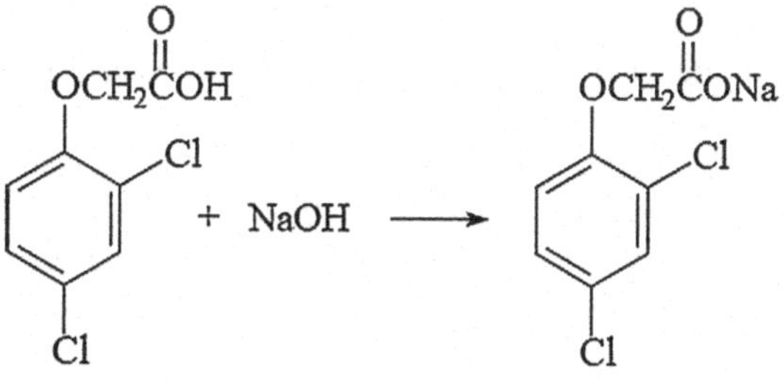

FIGURE 6.6 Sodium salt of 2,4-dichlorophenoxyacetic acid (2,4-D).

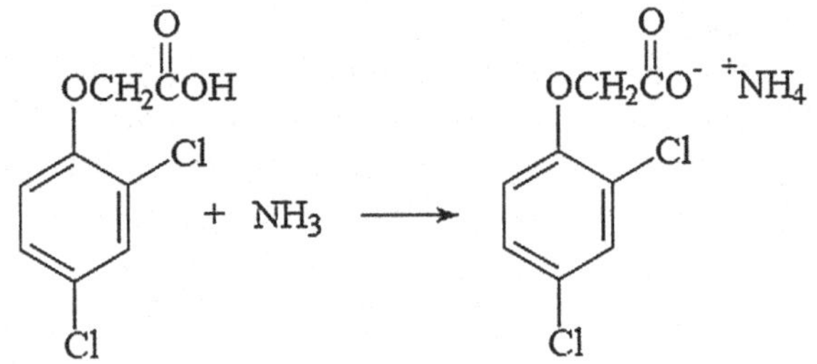

FIGURE 6.7 Amine salt of 2,4-dichlorophenoxyacetic acid (2,4-D).

emulsifying agent. At equal doses, they are more toxic because they are more readily absorbed by plant cuticle and cell membranes. The methyl, ethyl, and isopropyl esters are no longer commercially available because of their high volatility. The butoxyethyl ester and propylene glycol butyl ether ester have low volatility and thereby reduce, but do not eliminate, volatile movement.

Symptoms

External symptoms include epinastic (twisting and bending) responses, stem swelling and splitting, brittleness, short (often swollen) roots, adventitious[4] root formation, and deformed leaves. Symptoms often appear within hours of application—nearly always within a day. Epinastic responses result from differential growth of petioles and elongating stems. Leaf and stem thickening, leading to increased brittleness, often appear quickly. Color changes, cessation of growth, and sublethal responses occur. Plants often produce tumor-like proliferations and excessive adventitious roots. The effective dose varies with each weed species, its stage of growth at application, and the formulation applied. As plants mature, they can still be controlled by growth regulator herbicides, but higher rates are required. All or a few of these symptoms may appear in particular plants and activity is often due to two or more actions at the same time.

(4-Chloro-2-Methylphenoxy) Acetic Acid and 2,4,5-T

MCPA (Figure 6.8), developed in England, differs from 2,4-D by the substitution of a methyl (CH_3) group for chlorine at the two position of the benzene ring. Uses are similar and selectivity and effects are nearly identical. MCPA is more selective than 2,4-D in oats, but less 2,4-D is required to control many annual weeds. MCPA persists 2–3 months in soil, whereas 2,4-D persists about 1 month. Formulations are the same. MCPA is used more in the United Kingdom and in Europe than 2,4-D. MCPA is used in peas and flax in the United States because they are more susceptible to injury from 2,4-D.

---

4. Adventitious: structures that develop in an unusual place (e.g., above ground root formation).

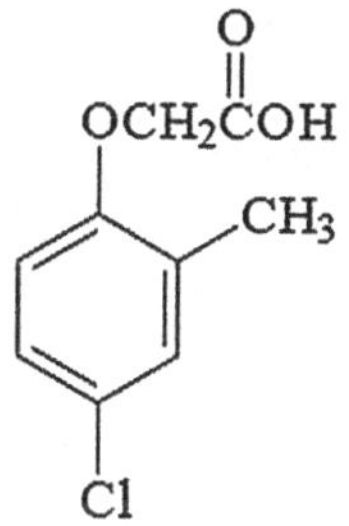

$$O$$
$$\parallel$$
$$OCH_2COH$$

FIGURE 6.8   MCPA: 2-methyl,4-chloro phenoxyacetic acid.

2,4,5-T (Figure 6.2) is no longer available in the United States. It is more effective against woody plants than 2,4-D. It was developed for control of brushy weeds and trees on rangeland. It was formulated as an amine or ester. It is more persistent than 2,4-D or MCPA. For several years, it was marketed in combination with 2,4-D for broad-spectrum control of broad-leaved weeds. It has low mammalian toxicity (50% lethal dose = 300 mg/kg; rat), but is no longer registered (approved for use) because of negative human health effects (see the section Agent Orange) that became painfully apparent during its military use for defoliation in Vietnam.

It didn't take long after activity was found with an acetic acid derivative of the phenoxy acid herbicides for researchers to examine the activity of the propionic (3 carbon), butyric (4 carbon), pentyl (5 carbon), or longer chain derivatives. Very early in the development of these compounds, it was found that a chain with an even number of carbons had herbicidal activity but a chain with an odd number did not. The even number carbon chain is broken down through beta oxidation (microbial cleavage of 2 carbon units) to produce 2,4-D, MCPA, or the appropriate analog with a 2 carbon chain. A 3, 5, 7, etc. carbon chain will also be broken down by beta-oxidation, but the final product is an alcohol that has no herbicidal activity. Thus, it is only the even-numbered carbon chains that have herbicidal activity. However, as is true for many generalizations about herbicides, this one is wrong. Straight chains follow the rule but iso- or branched chains do not. The alphaphenoxypropionic acids were widely used in Europe for weed control in small grains. Their structure has 3 carbons in a branched chain, which degrades like a 2 carbon chain. These compounds are dichloroprop (the analog of 2,4-D) or mecoprop (the analog of MCPA). Mecoprop was introduced in Europe as a complement to MCPA because of its ability to control catchweed bedstraw and common chickweed. Previously, these weeds could only be partially controlled by sulfuric acid or substituted phenolic acids.

Another part of the history of phenoxy acid herbicides is the phenoxybutyrics. MCPB and 2,4-DB—or 4-(2,4-DB)—were, but are no longer, used widely. MCPB is selective postemergence for annual broad-leaved weed control in peanut, soybean, and seedling forage legumes. Plants, through their enzyme composition, determine selectivity of 2,4-DB. Young alfalfa is less susceptible than older alfalfa, because older plants have a more efficient and widespread beta-oxidation system and are able to break down 2,4-DB to 2,4-D, which is immediately toxic.

MODERN WEED MANAGEMENT

Understanding the history, nature, properties, effects, and uses of herbicides is essential if one is to be conversant with modern weed management. Weed management is not accomplished exclusively with herbicides, but they dominate in the developed world. To most weed scientists, they are essential tools. Whether one likes them or deplores them, they cannot be ignored. To ignore them is to be unaware of the opportunities and problems of modern weed management. Ignoring or dismissing herbicides may lead to an inability to solve weed problems in many agricultural systems and may prolong development of better weed management systems. The majority of weed scientists think carefully about the link between what they do, the problems their work will solve, and the benefits. They are guided by clear utilitarian goals—to provide the greatest good for the greatest number of people. Theirs is a logical reason for doing weed science. It is science driven by the desirable goal of solving real problems rather than the more traditional scientific goal of intellectual curiosity (see Specter, 2007). In earlier times, "among elite scientists, it was usually considered gauche to be obsessed with anything tangible or immediate; brilliant discoveries were supposed to percolate" (Specter). They were rarely intentional. In the eighteenth and nineteenth centuries, that was how scientific research was done. It was the norm. In agricultural science most agricultural pest problems could not be solved, they could only be studied. That is no longer true.

Herbicide Development

A wide range of methods has been offered for vegetation management through the use of herbicides. Herbicide is derived from the Latin *herba* (plant) and *caedere* (to kill). Herbicides are chemicals that, when applied at the proper time, the proper stage of plant growth, and the proper dose, kill or reduce the growth and competitiveness of weeds. The definition accepted by the Weed Science Society of America (Vencill, 2002, p. 459) is that an herbicide is "a chemical substance or cultured organism used to kill or suppress the growth of plants." In effect, a herbicide disrupts the physiology of a plant over a long enough time to kill it or severely reduce its growth.

Pesticides are chemicals used to control pests. Herbicides differ from other pesticides because their sphere of influence extends beyond their ability to kill or control plants. Herbicides change the chemical environment of plants, which can be more easily manipulated than the climatic, edaphic (soil), or biotic environments. Herbicides reduce or eliminate labor and machine requirements and modify crop production techniques. When used appropriately, they are production tools that increase farm efficiency and may reduce energy requirements. Herbicides do not eliminate the requirement for petroleum energy because they are petroleum based.

Development of Weed Science

Although many tried, often with some success, to control weeds in the early twentieth century, weed science did not begin until the mid-twentieth century when weeds could be controlled selectively, quickly, at low cost with newly discovered herbicides. The Weed Science Society of America was founded in 1956. From the beginning, the purpose of weed science was to solve weed problems, primarily those in production agriculture. Weed scientists were in the business of reshaping how agriculture was practiced. Alden Crafts (1960) of the University of California at Davis, as he did often and well, told weed scientists in his presidential address to the Weed Society of America,[5] what their mission was. Primarily because of monocultural agriculture, "farmers are at war with weeds, the invaders of his crops." He continued, "at last man has devised tools for combating weeds, commensurate with the tools he uses for mining and manufacture and travel: modern mechanical and chemical tools." These new products contributed to the chemicalization of agriculture and are the "tools of the present day weed researcher." In his speech, Crafts reviewed the early discovery of the inorganic chemicals that were used for other purposes, but careful observers noted the death of weeds. He pointed out that although the discoveries were apparently accidental, "they had to happen." He saw the development of herbicides as an inevitable outcome of progress in plant physiology. In his view, the chemical control of weeds did not begin with the discovery of 2,4-D. It was a concept that "had to be born" because of accumulated knowledge in plant physiology, plant biochemistry, and the mechanisms of action of plant hormones.

Others agreed with Craft's idea of the inevitability of progress in weed management. For example, the North Central Weed Control Conference was created in 1944 by agronomists and weed scientists who came together with the common goal of discovering more effective control, including chemical control, of deep-rooted, noxious, perennial weeds, especially field bindweed. It was not the postwar availability of 2,4-D, but the concern about perennial weeds that moved scientists and administrators in the 14-state North Central region to confer. Work on 2,4-D was important, but it was not the reason the conference was created.

The first meeting of the Western Weed Control Conference was held in Denver in June 1938 (Appleby, 1993) well before the herbicidal activity of 2,4-D was discovered. The purpose of the conference was to foster other regional and a national weed control organizations. As Crafts (1960) predicted, progress in weed management had to happen.

---

5. The Weed Society of America was founded in 1956. The name was changed to Weed Science Society in 1967.

Crafts began a new science focused on weeds at the University of California at Davis. He said, "Little did I realize when on July 1, 1931, I Initiated weed control by scientific methods, that I was starting a technology that, in a mere 50 years would develop into an industry involving hundreds of effective herbicides that would exceed in cost and magnitude the sum total of all other pesticides" (Crafts, 1985; Shaw, 1984). Crafts and his mentor and colleague W.W. Robbins were among the creators of the scientific study of weeds. For them it was, because it had to be, a study of weeds rather than their control because there was no significant ability to control weeds. They were men born in the nineteenth century and educated in the early twentieth century. In 1910, feeding the horses and mules required to do the work on farms required more than one-fourth of the output of the world's farms and probably one-tenth of the required work on farms was devoted to caring for the draft animals. In 1922, a team of 32 mules was required in the state of Washington and much of the western United States for wheat harvest. It took one 32-mule team, 1 month to harvest 1200 acres of wheat. In the 1990s, a gas-powered combine completed the same harvest in a third of the time (Singer, 1998, p. 369) and it is accomplished faster now. Combine harvesters and other technical and mechanical developments, including pesticides were part of what has to be regarded as a revolution in the growing of food.

It is undeniable that weeds had been of concern to farmers and a few scientists long before 2,4-D and other herbicides appeared. There were "three full-time weed men (in the US) in 1934 and not too many part-time ones" (Willard, 1951). Oregon appointed the first full-time weed specialist in 1936 (Dunham, 1973). By 1951, the US Department of Agriculture (USDA) had the equivalent of 17 scientists working on weeds. In 1960, there were 66. Buchholz (1961) estimated that US states employed only 30 scientists who worked on weeds and most of them were part-time. By 1967, Buchholz estimated there were 160 state weed workers, whereas Dunham estimated, for the same time, that 17 states that had 20 full-time weed specialists and 89 specialists devoted part-time to weeds. One wonders why each of these men (they were all men) began to work on weeds. What drew them to weeds? Most were trained as agricultural scientists, but few had been involved in educational programs that produced weed scientists—those whose education and training focused on weeds. The dilemma of developing a discipline that claimed to be an objective science based on the study of a subjective class of plants is clear, but has not been questioned (Evans, 2002, p. 13). Weeds were defined and redefined and, although each weed scientist has a clear understanding of the objects of study, there is no universal definition, shared by all scientists. In 1967, the Weed Science Society of America defined a weed as a plant growing where it is not desired (Buchholtz, 1967). In 1989, the Society's definition was changed to "any plant that is objectionable or interferes with the activities or welfare of man" (Humburg, 1989, p. 267; Vencill, 2002, p. 462). The European Weed Research Society (1986) defined a weed as "any plant or vegetation, excluding fungi, interfering with the objectives or requirements of people." (For other definitions see Zimdahl (2007, pp. 17, 18).

Each definition is clear and each leaves the burden and the responsibility for specific identification and final definition with individuals. It is the individual who determines when a particular plant is growing in a place where it is not desired or when it interferes with their activities or welfare.

What was implicit but never made explicit in any definition or in the minds of weed scientists was that weeds are products of the way agriculture is practiced. They are products of ecology, psychology, and the culture that informs how we practice and think about agriculture (Evans, 2002, p. 14). This has not been considered carefully by weed scientists. If it is true, then the weed problem may be best addressed "by considering not only the agro-ecosystems that produce them but also the culture that informs how we farm and think about agriculture" (Evans, p. 14). Since the beginning of attention to weeds and especially since the advent of 2,4-D and its derivatives, weed scientists have been fully occupied with the multiple tasks defined by Willard (1951) and never abandoned. Willard asked "Where do we go from here?" His answer was "everywhere"; he was quite serious. As a developing science with a brand new technology whose actions and effects were only partially understood, literally everything was unknown and while Willard did not encourage "riding off in all directions," he did encourage building "a science and art which will go far to relieve the primeval curse placed on Adam and Eve when they were cast out of the Garden of Eden."

Weed scientists did not pause to examine the reasons weeds were omnipresent and seemed to become worse problems as production increased with new rapidly emerging technologies (e.g., fertilizer, new cultivars, pesticides, irrigation). Weed problems were real and increasing production was deemed to be an essential agricultural goal. The problem of weeds had to be solved and all engaged in agriculture adopted and were not often criticized for adopting what Evans (2002, p. 51) calls "a harsh, and at times blindly oppositional attitude" toward weeds. The search for cost-effective, efficient solutions to the omnipresent and worsening weed problems in existing systems of agriculture was the primary and worthy focus for weed scientists. Theirs were not systemic questions about the way agriculture was practiced. The primary questions and goals were directed toward controlling weeds to maximize crop production.

Farming systems changed slowly usually in response to other technological advances (machinery, improved cultivars, irrigation). Herbicides are the major technology whose use has been and is still investigated intensively by weed scientists. An exploration of the history of weed science (Zimdahl, 2010), which is beyond the scope of this book, must, in my view, include the stories of how herbicides were developed and their influence on the development of weed science. A few of the stories follow. Many of the stories of the development of important herbicide groups that ought to be included (e.g., Protox inhibitors, aryloxy-phenoxypropionic acids and cyclohexanediones) have not been told because those who were involved have died and their story died with them or the stories have not been written by those involved. A brief history of the development of several important herbicide chemical groups can be found in the introductory comments on the chemical groups in the three volume series edited by Kearney and Kaufmann (1975).

DEVELOPMENT OF 2,4-D

The phenoxy acid 2,4-D story begins, as so many scientific tales do, with discoveries made much earlier before anyone had discovered selective herbicides that could be applied at low doses. Those who were concerned about weeds surely had thought of and hoped for the possibility, but the basic research had not been done. As that work progressed, it may have appeared that the discoveries were accidental, but as mentioned above, in Crafts' (1960) view, "they had to happen." Crafts thought that the development of herbicides was an inevitable outcome of progress in plant physiology.

Peterson (1967) provides a clear description of the work that substantiates Crafts' claim that things like 2,4-D "had to happen." Peterson notes that "between 1880 and the mid-1930s, several botanists pursued different lines of investigation that made possible the discovery of 2,4-D." A few of the necessary antecedents noted by Peterson (1967) are repeated in the following sections[6] (readers are referred to his article for detailed citations).

The commercial history of 2,4-D can be traced to the first patents. US patents 2,322,760 (June 29, 1943) and 2,322,761 (June 29, 1943) for plant growth regulators were issued to John Lontz and assigned to E. I. du Pont de Nemours and Co. The Lontz patents address how growth characteristics of plants are modified by application of the compounds identified in the patent application.

On December 11, 1945, patent 2,390,941 was issued to Franklin D. Jones, who assigned it to the American Chemical Paint Co.[7] The Jones patent was for use. It described 2,4-D's uses as a weed killer, but it did not prevent another company from manufacturing and selling 2,4-D. The primary object of the "invention" was "to improve chemical methods for eradication of weeds in an active stage of growth." The secondary objective of Jones' patent was "to provide a wholly new class of systemic, translocated herbicides." Jones was studying ways to kill poison ivy motivated not only by commercial interests, but also by the fact that his children were highly sensitive to it. He began studying, as many others did, the plant growth hormone IAA, a naturally occurring compound found in most plants. It proved to be far too unstable to work effectively. His research led him to 2,4-D, a synthetic chemical that essentially did what IAA did.

---

6. Although not regularly cited, Peterson (1967) is the primary source for this section.

7. The American Chemical Paint Company was founded in Philadelphia in 1914 and relocated to Ambler, PA, in 1923 by James H. Gravell, an engineer and philanthropist. He bought the rights to Deoxidine, a chemical treatment for preventing rust on painted metal, which he had helped develop, from his employer (Hale-Kilburn Metal Co.). The American Chemical Paint Company became AmChem Products Inc. in 1959 and, after subsequent mergers, is now part of Aventis Crop Science.

He discovered that 2,4-D didn't increase plant development as expected. At higher doses, it killed broadleaf plants (dicots) with no adverse consequences for grasses and other monocots. Although 2,4-D did not succeed primarily as a plant growth regulator, its introduction as a herbicide (and sometimes a plant growth regulator) has increased world food production by helping to control weeds and other broadleaf plants. The discovery also laid the foundation for a tremendous amount of research on weed control, herbicide toxicity, and in the environmental sciences.

After the Jones patent was issued, 2,4-D was introduced commercially in 1946, and rapidly became the most widely used herbicide in the United States. 2,4-D has provided economical, effective, postemergence control of broadleaf weeds in a large variety of crops and in non-cropland for over the past six decades. Control of broadleaf weeds in turf grass was recognized as an important benefit by golf course superintendents and grounds managers as early as 1952. 2,4-D is still the third most widely used herbicide in the United States and Canada, and the most widely used worldwide. It is reasonable to claim as Hilton (2007) did that 2,4-D was essential to the development of the modern system of highly productive agriculture.

The discovery and development of the phenoxy acid herbicides is not the only interesting story of herbicide development, but they were the first and the story has not always been reported accurately. There are three important historical references, which have most of the details. The entire story will not be repeated here but some clarification is appropriate. The first report Peterson (1967) is complete about what happened in the United States, but ignores most of the very similar work that occurred in the United Kingdom at the same time. The second report (Kirby, 1980) was published by the British Crop Protection Council and understandably emphasizes the British work. Kirby, in her chapter on the American contribution, quotes extensively from Peterson's (1967) paper, although she does not cite his work. The discoveries in the United States and United Kingdom were not the beginning of attempts to control weeds with chemicals but they were the beginning of modern chemical weed control. All previous herbicides were just a prologue to the rapid development that occurred after the discovery of the selective activity of the phenoxyacetic acid herbicides. Readers interested in more detail are encouraged to consult Peterson (1967), Kirby (1980), and, for the most complete story, Troyer (2001). Troyer, a botanist from North Carolina State University, wrote the most complete and accurate historical report. He correctly claims that there were a multiple independent discoveries of the phenoxy acids by four groups of workers in two countries.

Herbicide development programs expanded rapidly after 2,4-D appeared. The American Chemical Paint Company was one of a very few companies involved in herbicide development. Several companies manufactured and sold chemicals (e.g., acrolein, allyl alcohol, arsenic acid, sodium chlorate, sodium arsenite, and Stoddard solvent [a hydrocarbon mixture]) that were moderately effective herbicides. These companies were not engaged in what soon became

TABLE 6.1 Annual Herbicide use in the United States

Year	Pounds of Herbicides Used
1950s	14 million
1960s	50 million (Canine, 1995)
1995	56 million (574 different products; Canine, p. 201)
2000	542 million (cost roughly $6.4 billion)
2007	531 million (cost roughly $5.9 billion)

the new industry of discovery and development of herbicide chemistry. Companies did not immediately move into this new business for several reasons. Perhaps the most important was that no one knew how to find a new herbicide or any other pesticide. Pharmaceutical companies had screening methods to find new drugs, but similar methods had not been developed to find new pesticides. The search was going to be expensive and those who ran companies surely wanted to be convinced that the necessary investments would result in profit. Finally, there was uncertainty that farmers would adopt the new technology. But, 2,4-D's immediate and continued success allayed fears about farmer acceptance. "In 1945, the first year 2,4-D was sold to the public, total production in the United States was 917,000 pounds (Canine, 1995, p. 201). In 1946, 5.5 million pounds were produced. Annual use of herbicides has increased steadily (Table 6.1).

About 5.2 billion pounds of all pesticides were used in the world in 2006 and 2007; herbicides were 40% (2 + billion pounds) of the world total.[8] Between 1980 and 2007, the aggregate quantity of pesticides applied in the United States declined at an average rate of 0.6%/year, whereas inflation-adjusted expenditures increased 0.6%/year. The prices paid by agricultural producers for fuel, seed, fertilizer, and labor increased roughly twice as fast as the prices paid for pesticides during this period. Herbicides accounted for 47% of the pesticides applied in the United States and 25% of world use in 2007.

The great era of herbicide development came at a time when world agriculture was involved in a revolution of labor reduction, increased mechanization, and new methods to improve crop quality and produce higher yields at reduced cost. Herbicide development was an integral part of developed country agriculture's entry into the chemical era (see Chapter 2). It included reduced need for human labor, increased mechanization, and new technology that improved crop quality and produced higher yields at reduced cost. Herbicide development built on past technological improvements and, with the development of insecticides

---

8. http://www.epa.gov/opp00001/pestsales/07pestsales/usage2007.htm#2_1. Accessed August 2014.

and fungicides, began the transformation of farming from a mechanical, labor intensive activity to an energy dependent, chemical-based system of food production.

Farmers were ready for improved methods of selective chemical weed control. Their acceptance of technological developments that changed the practice of agriculture has been characterized in terms of economic, political, social, and philosophical attitudes by Perkins and Holochuck (1993). Farmers wanted to improve their operation in competition with other farmers and were willing to adopt new technology that enabled them to improve their *economic* competitiveness. New technology was *socially* acceptable because as independent entrepreneurs many technological innovations could be used to gain advantage, independent of neighbors. *Politically*, farmers welcomed technical assistance that came from public laboratories and land-grant universities and government price support systems that allowed farm operations to remain private. Farmers were highly social people who remained fiercely independent and welcomed opportunities to do what they wanted on their farms. New technology—developed at no apparent cost to the farmer—that could be adopted without interference from anyone was welcomed. Finally, *philosophically*, farmers perceived that a major part of farming was controlling nature; bending it to human will. Although this was a never-ending challenge, success was apparent when technology that increased production was readily available. Herbicides fit well in each category.

It is true that no weed control method has ever been abandoned; new ones have been added and the relative importance of methods has changed. The need for cultivation, hoeing, etc. has not disappeared. These methods persist in small scale and developing country agriculture. Older methods have become less important in developed country agriculture because of the rising costs of labor, the availability of effective chemical methods, and narrower profit margins.

A 1945 survey of commercially available herbicides in the United States (Kephart, 1947) documented 51 different products. Of those, 25 contained arsenic, five incorporated either nitrophenol or sodium chlorate, three were phenoxyacetic acids, and a few contained boron or copper salts. Others were based on various inorganic materials with herbicidal activity. Petroleum-based herbicides, first used in California on non-cropland in 1924, were widely used by 1935 in southeastern states. In the early 1940s, petroleum oils were used for selective weed control in carrots (Dunham, 1973). The Ashgate handbook (Milne, 2000) includes all types of pesticides and other chemicals used in agriculture, including fertilizers. Herbicides are the largest single group with 835 named products manufactured from 135 different chemical compounds. Eighty-one of the 835 contain some percentage of 2,4-D. If proprietary labels are considered, there may be more than 1000 chemical and biological compounds used for pest control in the world (Hopkins, 1994).

Rapid development of herbicides occurred after WW II. Shaw (1954) discussed the scope of chemical weed control in the United States. He reported on six important classes of herbicides (phenoxy and phenoxypropionic acids, benzoic acids, substituted phenols, carbamates, substituted ureas, and a few diverse chemical structures) that were in use or being developed. By 1954, most inorganic herbicides were no longer widely used. In 1954, Shaw estimated that the then-huge amount of 85 million pounds of herbicides was used annually in the United States. One of every 10 US crop acres was treated with a herbicide. In 2002, 204 selective herbicides were listed in the Weed Science Society of America's (WSSA) *Herbicide Handbook* (Vencill, 2002) and by 2004, 357 had been approved by the WSSA (Anonymous, 2004). In addition, there were several experimental herbicides in some stage of progress toward marketability. If dollars of product sold are the criterion used, pesticide use has been increasing and herbicide use dominates. In 1997, 1 billion pounds of pesticides were used in the United States and more than 47% (461 million pounds) were herbicides (Gianessi and Silvers, 2000). Just 10 herbicides accounted for 75% of sales (Gianessi and Marcelli, 2000). US farmers routinely apply herbicides on more than 85% of crop acres (Gianessi and Sankula, 2003). A study of 40 crops showed treatment of 220 million acres at a cost of $6.6 billion (Gianessi and Sankula). In 2001, the global market for nonagricultural pesticides was more than US $7 billion/year and was growing about 4% a year. The global market just for turf pesticides is approximately $850 million/year; about half is used on golf courses. Each year, US lawn care firms apply about $440 million worth of pesticides.

The National Agricultural Statistics service of the USDA regularly surveys selected states and selected crops to determine the extent of fertilizer and pesticide use. Reports are available for 1990 through the current decade at http://usda.mannlib.cornell.edu (accessed January 2008; enter herbicides in the search box and click on Agricultural Chemical Usage–Chemical Distribution Rate and select the category of interest).

Companies and Sales

The global herbicide market was estimated to be $13.5 billion from 1990 to 1993 and one-third ($4.5 billion) was the US market. Kiely et al. (2004) estimated that $14 billion was spent worldwide on chemical weed control. Japan was the next largest single country, with $1.5 billion in sales. When the entire European market was considered, it was second largest, with France ($1.25 billion) the largest single European user (Hopkins, 1994). In 2001, world expenditures on all herbicides were US $14,118 million. US users spent $6410 million for 553 million pounds of active ingredient, which was equal to 4987 million pounds of product. These amounts are lower than purchases in 2000 and use has returned to the levels last seen in the early 1970s (US/EPA, 2004). Of these amounts, 78% is used in agriculture, with the rest nearly evenly divided between industrial/commercial/government (12%) and home and garden use (10%).

In 1990, about 45% of world pesticide sales volume was herbicides (similar to the US data), insecticides were 28%, and fungicides approximately 20% of total sales (Hopkins, 1994). More than 85% of herbicides are used in agriculture.

The worldwide market is becoming increasingly concentrated in the hands of a few multinational corporations engaged in herbicide discovery, development, and sales. Nearly half of the companies in pesticide discovery (but not in development and marketing) in 1994 were Japanese (Hopkins). The number of companies marketing herbicides in the United States has steadily declined from 46 in 1970 to 7 in 2005 (Appleby, 2006). Three are based in the United States; the others are in Europe. All operate in the United States (Appleby, 2006, personal communication).

Worldwide sales have continued to increase. World exports of pesticides of all kinds totaled $15.9 billion in 2004, a new high in sales for the global chemical industry (Jordan, 2006). Use of all kinds of pesticides has risen from nearly 0.5 kg/hectare in 1960 to 2 kg/hectare in 2004. The recent increase is attributed mainly to the increased use of herbicides on genetically modified crops in China (Jordan).

Timmons (1970) reported 75 herbicides marketed between 1950 and 1969. Appleby (2006) included 184 herbicides marketed between 1970 and 2005, a 2.4 times increase. The number of herbicides listed in the first (1967) through the eighth (2002) edition of the Weed Science Society of America's Herbicide Handbook has increased, as has the number of different chemical families in which herbicidal activity has been discovered. Similarly, the number of WSSA-approved herbicides has increased from 304 in 1995 (Anonymous, 1995) to 357 in 2004 (Anonymous, 2004). It is clear that, because of their significant production advantages and efficiency, herbicides dominate modern weed control. Although the herbicide chemical industry has undergone extensive consolidation, as have many other manufacturing industries, that has not diminished discovery and development of new herbicides in older chemical families or discovery of activity in new chemical groups.

"Pesticides—whether insecticide, herbicide or fungicide—vary in their effectiveness at killing their specific targets. But a property pesticide chemicals share is their ability to trigger strong emotions" (Wickens, 2002). The first serious challenge to widespread use, misuse, and the lack of adequate regulation of pesticides was Carson's (1962) book *Silent Spring* (see Chapter 7). Criticism continues (Bittman, 2013), although debate and resolution are infrequent. Carson's book is widely and correctly credited with helping launch the American environmental movement. The New Yorker started serializing *Silent Spring* in June 1962. The book was published by Houghton Mifflin in September 1962. *Silent Spring* and the environmental movement (the greening of America) have had major effects on the pesticide industry and pesticide use.

WORLD WAR II

Chemical and commercial history are important parts of the phenoxy acid story. Another part is the many related events and the good scientists whose work was affected by war. As mentioned previously, none of the scientific research in the early 1940s proceeded toward normally expected publication in the United States or United Kingdom because of financial and security restrictions imposed by World War II. Wartime secrecy mandated that the work could not be reported in an accepted, timely manner in the scientific literature. Thus, the determination of who was first to develop and recognize the potential of the phenoxy acetic acids as selective herbicides and their acceptance by farmers could only be determined when the research was finally published, often years after it had been done.

Four Concerns

The war raised three concerns. The first was that, although the biological (the herbicidal) potential of the new chemicals was beginning to be understood, it was not understood well enough to know if they could be used as weapons of war to destroy enemy crops. The decision-makers in both countries recognized the potential of biological warfare to destroy crops but neither group was convinced it could be done or, perhaps more importantly, that it should be done. The first, a scientific/technical question, asked if there was the equipment (the mechanical (sprayer) technology and the aerial technology (enough planes)) to do what would have to be done. The second was a moral question. Should we do it even if we can? It was the more difficult question. WW II was not dominated by moral decisions. The ultimate answer in both countries was "no" to both questions. One cannot know at this juncture what the answer to the moral question would have been if the technical issues had been resolved. Kirby (1980, p. 7) suggests that both countries "eagerly seized" upon use of the chemicals (2,4-D was a primary candidate) for destroying enemy crops only to abandon it at the last moment because the chemicals were considered too specific for the plants they killed and the techniques (of application) would be more costly than the high density bombing of German towns then under way." Third, there was a real fear that German chemists might have made the same discoveries and if the allies did not act the Luftwaffe would. The British, especially, wanted to know what would happen if the Germans were to use such chemicals on their crops. Finally, the dominant question was not what the effects might be. They were known. The question was, if the Germans had and used the chemicals, were there solutions or would all crops die and people starve? Herbicides were not used in World War II, but as a result of a suggestion by Ezra Kraus, use was strongly considered.

Ezra J. Kraus

Ezra J. Kraus began his career as an assistant in horticulture at Oregon State College in 1909. His doctoral degree was completed at the University of Chicago in 1917 while on leave from Oregon. He became dean of the Service Departments at Oregon State College (art and rural architecture, bacteriology, botany, chemistry, English, entomology, history, mathematics, modern language, physics, public speaking, zoology, and physiology). One wonders exactly how he had time to serve anyone. Kraus left Oregon in 1919 for a position in applied botany at the

University of Wisconsin. Subsequently, he moved to the University of Chicago as head of the Department of Botany. He was awarded an honorary Doctor of Science degree by Oregon State College in 1938 and the same degree by Michigan State University in 1949. He died in Corvallis, OR, in 1960.

He studied plant growth regulation for several years and much of his work is regarded as foundational to the field of plant (crop) physiology. He supervised the doctoral programs of J.W. Mitchell and C.L. Hamner who, in the early 1940s, were studying the physiology and activity of growth regulators at the USDA, Plant Industry Station at Beltsville, MD. Kraus is relevant to the 2,4-D story because he knew that plant growth regulators distorted plant growth if used at higher than growth-regulating doses and might be used beneficially to selectively kill plants. He saw potential use as chemical plant killers (herbicides) and advocated purposeful application in toxic doses to kill the German potato crop and thus hasten the end of the war. He and other prominent scientists convinced then US Secretary of War H.L. Stimson that others might develop and use the chemicals for biological warfare against the United States. They advocated that US scientists begin research to determine if there was potential to use the growth regulating chemical (e.g., 2,4-D) against German crops. The work was done under contract from the US Army at its Biological Warfare Laboratory at Camp (Fort) Detrick, near Frederick, MD (Peterson, 1967; Troyer, 2001). Hamner and Tukey (1944a,b) reported the first field trials with 2,4-D for successful selective control of broad-leaved weeds. They also worked with 2,4,5-T as a brush killer. Marth and Mitchell (1944), also Kraus's students, first reported the differential use of 2,4-D for killing dandelions and other broad-leaved weeds selectively in Kentucky bluegrass turf. Marth and Mitchell attributed the quest for selective activity of the growth regulator chemicals to Kraus.

Herbicides had been used in Malaysia, a British colony, after 1873. The British used herbicides for defoliation and crop destruction between 1951 and 1953 in their armed struggle against Malaysian communist guerilla insurgents. British interests included protecting their tin mining and rubber plantation businesses. The United States used massive doses of phenoxy acid and other herbicides in Vietnam for several years after 1962 (Troyer, 2001; see his references for more details). Peterson (1967) notes that the herbicides developed under wartime secrecy for military uses in the early 1940s were used by the US military in Vietnam.

Agent Orange

The discovery and development of 2,4-D, and 2,4,5-T have been described previously, but there is another story that must be included. Phenoxy acids were studied during World War II because they had potential as weapons of biological warfare against German crops. They were not used as biological weapons by the United States until January 13, 1962, when three US Air Force C-123 planes took off from Tan Son Nhut Airbase in what was then South Vietnam to begin Operation Ranch Hand. The C-123 was a twin-engine, propeller-driven, short-range assault and transport plane. Each plane was loaded with more than 1000 gallons of one of eight herbicide formulations. Agent Orange had four formulations each with esters and different percentages of 2,4-D and 2,4,5-T. After 1968, the dominant formulation was as an equal combination of the *n*-butyl ester of 2,4-D and the isooctyl ester of 2,4,5-T. Agent Orange was one of the rainbow agents (blue, green, orange, pink, purple, and white) that had been developed for use in South Vietnam after preliminary experiments by the USDA in Puerto Rico and at Camp Drum in New York state. Agent Orange became the best known of the herbicide mixtures. The recommended use rate in Vietnam was at least 10 times or more higher than normal field use rates of phenoxy acid herbicides in the United States.

Testing in Puerto Rico and South Vietnam revealed that the *n*-butyl ester of 2,4,5-T used in combination with the *n*-butyl ester of 2,4-D (used between 1965 and 1970) effectively eliminated nearly all unwanted broad-leaved plants. When 2,4,5-T is manufactured, temperature control is required to minimize formation of an undesirable, nonphytotoxic contaminant—2,3,7,8-tetrachloroparadioxin. It is one member of a family of compounds known as dioxins and is a potent teratogen. Approximately 65% of all herbicides used in Vietnam contained 2,4,5-T, which was contaminated with varying levels of dioxin (Stellman et al., 2003). A teratogen can cause terata or birth defects when pregnant women are exposed. Dioxins also cause chloracne, a skin condition characterized by blisters and irritation. There was never any debate about whether the dioxin contaminant in 2,4,5-T was a teratogen or caused chloracne. Most of the concern and debate ensued because of the unknown level of exposure and harm to Vietnamese citizens, Vietnam-era US military personnel, and pregnant women that was or might have been due to the dioxin contaminant.

In 1962, 15,000 gallons of Agent Orange were sprayed over Vietnam. By 1966, 2.28 million gallons were sprayed (Quick, 2008). The undeniable pursuit of biological warfare ended in 1970 because of increasing public and Congressional concern over the potential human and actual ecological dangers of the herbicides that by 1970 had been sprayed on a seventh of Vietnam's total land area. The phenoxy herbicides in Agent Orange surely had ecological effects and caused ecological harm. There is still vigorous debate about whether the herbicides as they were used in Vietnam harmed human physiology. It is certain that the dioxin contaminant has the potential to cause harm. There is little debated about the psychological harm Agent Orange has caused.

The official history of Operation Ranch Hand was written by James Clary, a US Air Force officer. Clary admitted in a Congressional hearing conducted by Senator Daschle of South Dakota that "When we initiated the herbicide program in the 1960s we were aware of the potential for damage due to dioxin. We were even aware that the military formulation had a higher dioxin concentration than the civilian version, due to lower cost, and the desired (if not mandated) speed of manufacture. However, because the material was to be used on the enemy, none of us were overly concerned" (Quick).

Clary's comment is, one assumes, typical of the attitude of military people toward whoever is defined as the enemy. His words are reflective of General US Grant's comments on the art of war.

The art of war is simple enough. Find out where your enemy is. Get at him as soon as you can. Strike him as hard as you can, and keep moving on.[9]

The aim of war is to win. Whether using Agent Orange and other herbicides as weapons of war was a good thing in Vietnam will be debated for many years. Their use in a war is a part of the phenoxy acid story. In April, 1970, 2,4,5-T was banned from most US domestic uses by the US Environmental Protection Agency. For a history of research on Agent Orange see Butler (2005).

EFFECT ON AGRICULTURE

The development of 2,4-D and the other phenoxyacetic acid herbicides quite literally transformed agriculture in much of the world and should be ranked as one of the great contributions of science. For the first time, farmers and others were able to control many broad-leaved weeds selectively and inexpensively in monocotyledon (grass) crops (e.g., wheat, corn) without obvious harm to them, other species, or the environment. The advent of 2,4-D (a term not used in the literature until 1945) (Peterson, 1967) revolutionized crop production and created the pesticide chemical industry (Fryer, 1980). The selective properties of the phenoxyacetic acids were revolutionary; the use rates were even more so. Before their development, it was necessary and acceptable to use the available inorganic herbicides at 75–150 pounds/acre. 2,4-D was quite amazingly effective and selective at a few pounds/acre or even less.

REFERENCES

Akamine, E.K., 1948. Plant growth regulators as selective herbicides. Hawaii Agricultural Experiment Station Circle 26, 1–43.

Anonymous, 1995. Common and chemical names of herbicides approved by the Weed Science Society of America. Weed Science 43, 328–336.

Anonymous, 2004. Common and chemical names of herbicides approved by the Weed Science Society of America. Weed Science 52, 1054–1060.

Appleby, A.P., 1993. The Western Society of Weed Science 1938–1992. The Western Society of Weed Science, Newark, CA. 177 pp.

Appleby, A.P., 2006. Weed Science Society of America - Origin and Evolution - the First 50 Years. Weed Sci. Soc. America, Lawrence, KS. 63 pp.

Bittman, M., January 4, 2013. Pesticides: Now More than Ever. NY Times Opinionator. Http://opinionater.blogs.nytimes.com/2012/12/11/pesticides-now-more-than-ever/?pagewant (accessed January 2013).

---

9. Brinton, J.H. 1914. Personal Memoirs of J.H. Brinton, Major and Surgeon, 1861–1865. Shawnee Classic Series. S. Illinois University Press. Carbondale, IL. 380 pp. Quote from U.S. Grant on p. 239.

Buchholz, K.P., 1967. Report of the terminology committee of the Weed Science Society of America. Weeds 15, 388–389.

Buchholz, K.P., 1961. Weed control – a record of achievement. Weeds 10, 167–170.

Butler, D.A., 2005. Connections: the early history of scientific and medical research on "agent orange". Journal of law and Policy 13 (2), 527–549 (Brooklyn Law College, Brooklyn, NY).

Canine, C., 1995. Dream Reaper - The Story of an old-Fashioned Inventor in the High-Tech, High-Stakes World of Modern Agriculture. A.A. Knopf, New York. 300 pp.

Carson, R., 1962. Silent Spring. Houghton Mifflin, New York.

Crafts, A.S., 1960. Weed control research—past, present, and future. Weeds 8, 535–540.

Crafts, A.S., 1985. Reviews of Weed Science - Dr. Alden S. Crafts. Reviews of Weed Science 1: iv.

Dunham, R.S., 1973. The Weed Story. Institute of Agriculture, University of Minnesota, St. Paul, MN. 86 pp.

Evans, C.L., 2002. The War on Weed in the Prairie West – an Environmental History. University of Calgary Press, Calgary, Alberta, Canada. 309 pp.

Fryer, J.D., 1980. Foreword. pp. 1–3. In: Kirby, C. (Ed.), The Hormone Weed Killers: A Short History of Their Discovery and Development. British Crop Protection Council Pubs, Croydon, UK, p. 55.

Gianessi, L.P., Marcelli, M.B., 2000. Pesticide Use in U.S. Crop Production: 1997. National Center for Food and Agricultural Policy, Washington, DC.

Gianessi, L.P., Sankula, S., 2003. The Value of Herbicides in U.S. Crop Production – Executive Summary. National Center for Food and Agricultural Policy, Washington, DC.

Gianessi, L.P., Silvers, C.S., 2000. Trends in Crop Pesticide Use: Comparing 1992 and 1997. National Center for Food and Agricultural Policy, Washington, DC. 165 pp.

Hamner, C.L., Tukey, H.B., 1944a. The herbicidal action of 2,4-dichlorophenoxy acetic and 2,4,5-trichlorophenoxyacetic acid on bindweed. Science 100, 154–155.

Hamner, C.L., Tukey, H.B., 1944b. Selective herbicidal action of midsummer and fall applications of 2,4-dichlorophenoxyacetic acid. Botanical Gazette 106, 232–245.

Hilton, J.L., 2007. History and Me. Self-published. 172 pp.

Hopkins, W.L., 1994. Global Herbicide Directory, first ed. Ag. Chem. Information Services, Indianapolis, IN. 181 pp.

Humburg, N.E. (Ed.), 1989. Herbicide Handbook, sixth ed. Weed Sci. Soc. Am., Champaign, IL. 301 pp.

Jordan, L.H., 2006. Pesticide trade shows new market trends. pp. 28–29. In: Assadourian, E. (Ed.), Vital Signs: The Trends that Are Shaping Our Future. W.W. Norton & Co, New York.

Kephart, L.W., 1947. Technical and commercial aspects of 2,4-D. Agriculture Chemicals 8 (25–27), 59–61.

Kiely, T., Donaldson, D., Grube, A., 2004. Pesticide Industry Sales and Usage: 2000 and 2001 Market Estimates. Environmental Protection Agency (EPA), Washington, DC.

Kirby, C., 1980. The Hormone Weed Killers: A Short History of Their Discovery and Development. British Crop Protection Council Pubs, Croydon, UK. 55 pp.

Kearney, P.C., Kaufman, D.D., 1975. Herbicides - Chemistry, Degradation, and Mode of Action. Marcel Dekker, Inc, New York. 500 pp.

King, L.J., 1966. Weeds of the World: Biology and Control. Interscience Pub., Inc, New York. 526 pp.

Kögl, F., Erxleben, H., Haagen-Smit, A.J., 1934. Über die Isolierung der Auxine a und b aus pflanzlichen Materialien. IX. Mitteilung. Physiological Chemistry 243, 209–226.

Marth, P.C., Mitchell, J.W., 1944. 2,4-dichlorophenoxyacetic acid as a differential herbicide. Botanical Gazette 106, 224–232.

Milne, G.W.A., 2000. Ashgate Handbook Pesticides and Agricultural Chemicals. Ashgate Publications, Ltd, Hampshire, England. 206 pp.

Norman, A.G., Minarik, C.E., Weintraub, R.L., 1950. Herbicides. Annual Review of Plant Physiology 1, 141–168.

Perkins, J.H., Holochuck, N.C., 1993. Pesticides: historical changes demand ethical choices. pp. 390–417. In: Pimentel, D., Lehman, H. (Eds.), The Pesticide Question: Environment, Economics, and Ethics. Chapman & Hall, New York. 441 pp.

Peterson, G.E., 1967. The discovery and development of 2,4-D. Agricultural History 41, 243–253.

Pokorny, R., 1941. Some chlorophenoxyacetic acids. Journal of the American Chemical Society 63, 1768.

Quick, B., 2008. The Boneyard—agent Orange a Chapter of History that Just Won't End. Orion. March/April, 16–23.

Shaw, W.C., 1954. Recent advances in weed control in the United States. In: Proc. British Weed Control Conference, pp. 23–46.

Shaw, W.C., 1984. Weed science: Revolution in agricultural technology. Weeds 12, 153–162.

Singer, E.N., 1998. 20th Century Revolutions in Technology. Nova Science Pub., Commack, NY (pp).

Skoog, F., Thimann, K.V., 1934. Further experiments on the inhibition of development of lateral buds by growth hormones. Proceedings of the National Academy of Sciences 480, 482–483.

Slade, R.E., Templeman, W.G., Sexton, W.A., 1945. Plant growth substances as selective weed killers. Nature (London) 155, 497–498.

Specter, M., 2007. Darwin's Surprise. The New Yorker. December 3, pp. 64–71.

Stellman, J.M., Stellman, S.D., Christian, R., Weber, T., Tomasallo, C., 2003. The extent and patterns of usage of agent orange and other herbicides in Vietnam. Nature 422, 680–686.

Templeman, W.G., 1939. The effects of some plant growth substances on dry-matter production in plants. Empire Journal of Experimental Agriculture 7, 76–88.

Timmons, F.L., 1970. A history of weed control in the United States and Canada. Weed Science 18, 294–307 Republished – Weed Science 53: 748–761.

Troyer, J.R., 2001. In the beginning: the multiple discovery of the first hormone herbicides. Weed Science 49, 290–297.

US/EPA, 2004. Pesticide Industry Sales and Usage: 2000 and 2001 Market Estimates. http://www.epa.gov/oppbead1/pestsales/index.htm.

Vencill, W.K. (Ed.), 2002. Herbicide Handbook, eighth ed. Weed Sci. Soc. Am., Lawrence, KS. 493 pp.

Wickens, B., 2002. The killing fields: pesticides work. to some folks that's the problem. August 19 Maclean's 35.

Willard, C.J., 1951. Where do we go from here? Weeds 1, 9–12.

Zimdahl, R.L., 2010. A History of Weed Science in the United States. Elsevier, London, UK. pp. 97–99.

Zimdahl, R.L., 2007. Fundamentals of Weed Science, third ed. Academic Press, San Diego, CA. 666 pp.

Zimmerman, P.W., Hitchcock, A.E., 1935. Several chemical growth substances which cause initiation of roots and other responses in plants. Contributions from Boyce Thompson Institute 7, 209–229.

Zimmerman, P.W., Hitchcock, A.E., 1942. Substituted phenoxy and benzoic acid growth substances and the relation of structure to physiological activity. Contributions from Boyce Thompson Institute 12, 321–343.

Chapter 7

DDT: An Insecticide

The early '60s ushered in an era revolutionary thinking—civil rights took the forefront, rock-n-roll seized the stage, and folks like John Coltrane and Andy Warhol reinterpreted everything from jazz to soup cans. In literature, Maurice Sendak described where the wild things were, Harper Lee encouraged us not to kill them, and Rachel Carson revealed how it was too late; we already were. For millions of maturing baby boomers, Carson turned cautionary eyes to another post-war product that was similarly coming of age—the organochlorine pesticide DDT.

Ziolkowski et al. (2010).

INTRODUCTION

When people hear the word "pesticide," it is commonly, immediately, and negatively associated with danger. If asked, many people will offer dichlorodi-phenyltrichloroethane (DDT) as an example of the worst of the entire class of pesticides. Because DDT is probably the most famous pesticide ever developed, it is not surprising that it is so quickly mentioned. It has been around for quite a long time. The intent of this chapter is to present enough of the several aspects of the DDT story to establish its relationship to and its role in the practice and transformation of agriculture. Readers who want a complete DDT story are referred to the excellent book by Kinkela (2011).

Le Couteur and Burreson (2003) point out the irony of the chlorocarbon pesticides. "Those that have done the most harm or have the potential to do the most harm seem also to have been the very ones responsible for the some of most beneficial advances in our society." The beneficial chlorocarbons include anesthetics and refrigerants that make food storage safe and air-conditioners possible. They make our drinking water safe and have eliminated or greatly reduced the scourge of insect-borne diseases, particularly malaria, which has killed more humans than any other disease.

Six Chemicals That Changed Agriculture. http://dx.doi.org/10.1016/B978-0-12-800561-3.00007-9

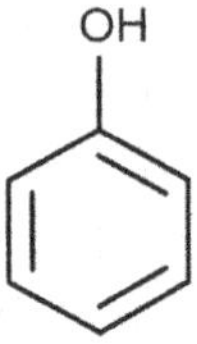

FIGURE 7.1   Phenol.

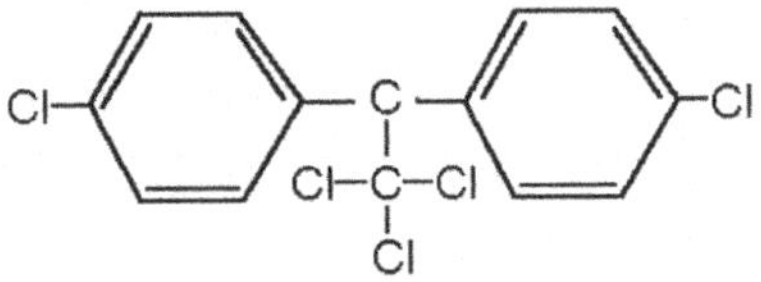

FIGURE 7.2   DDT.

HISTORY[1]

DDT is one member of an enormous group of chemicals, including 2,4-dichlorophenoxyacetic acid (2,4-D; Chapter 6) based on phenol (Figure 7.1). The group includes those that impart flavor (e.g., vanilla), explosives, pesticides, plastics, and structural materials (wood, lignin). DDT ($C_{14}H_9Cl_5$) (Figure 7.2) was first synthesized in 1873 by Othmar Zeidler, an Austrian student. Similar to Pokorny's synthesis of 2,4-D, Zeidler's work and DDT did not receive any particular attention. The insecticidal properties of multiple chlorinated aliphatic chemicals with at least one tri-chloro methyl group were described in 1934 by W. von Leuthold in Swiss patent #DRP Nr 67324627.4. After 4 years of intensive research on chlorinated aliphatics, Paul Müller, who was employed by J.R. Geigy A.G. of Basle, Switzerland, synthesized DDT. The basic Swiss patent was granted in 1940. Field trials showed it to be effective against the common housefly and a wide variety of insect pests, including the louse,[2] the Colorado potato beetle, and mosquitoes.[3]

Louse (lice) infestations have plagued soldiers and poor people for centuries. Concern about human louse-borne disease was so great that after the World War I armistice in 1918, returning troops were deloused at home ports and quarantined for 2 weeks (Graham, 1921). At the beginning of World War II, in full knowledge of the inevitable louse problem, soldiers were dusted with 96% naphthalene, 2% creosote, and 2% iodoform powder and their clothing seams were smeared with vermijelli (crude mineral oil, soft soap, and water). Until DDT appeared, MYL powder was the primary louse powder used by

---

1. The primary reference on history is http://en.wikipedia.org/wiki/DDT.
2. Louse (plural: lice) is the common name for more than 3000 species of wingless insects of the order *Phthiraptera*, three of which are human disease agents.
3. A family of nematocerid flies: the *Cuilicdae* with 41 genera and more than 3500 species.

the Army. It contained 0.2% pyrethrins, 2% *n*-isobutyl undecylenamide (to improve pyrethrin's efficacy), 0.25% Phenol S (to prevent deterioration of pyrethrins), and 2% 2,4-dinitroanisole (to kill louse eggs).[4] The powder was tested by the US Food and Drug Administration and found safe to use on man. Pyrethrins were the chemicals of choice for control of most insects until DDT came along.

In 1942, a team of US Department of Agriculture entomologists, led by W. E. Dove, was charged with developing a pesticide to prevent louse-borne typhus in troops. They tested thousands of chemicals and found at least one that killed lice. Geigy began manufacturing two products based on DDT, Gesarol and Neocide, in 1942. Dove's group received a waxy, granular sample, but the crumbly wax didn't work well as a delousing treatment, so they reformulated it. By 1943, Geigy was producing large quantities of several formulations of DDT, including a formulation of Gesarol from Geigy.[5] They found that Gesarol killed lice, and every other insect, in powders and sprays. Gesarol was then referred to by its generic name, abbreviated to DDT.

The US Department of Agriculture scientists promoted DDT only for a few circumscribed uses, including delousing and malaria control. Indeed, Dove specifically cautioned against spraying the stuff willy-nilly outdoors, arguing as early as 1944 that DDT was definitely poisonous and that its environmental consequences might be bad (see Footnote 5). His caution was that of a good scientist who knew that all the possible characteristics of this very effective insecticide were not known. However, it was used widely during World War II, with extensive human exposure. There was no immediate or subsequent evidence that it was harmful to people.

British and American medical entomologists learned of the products during World War II, when supplies of pyrethrum,[6] which had been used for centuries as an insecticide, were rapidly falling short of demand primarily for control of the female anopheles mosquitoes, the primary vector of malaria. Production was soon established on both sides of the Atlantic. DDT proved to be of inestimable value for control of disease-carrying insects during World War II and was a major factor in the success of several Allied war campaigns. Aerial spraying of DDT nearly eliminated the insect vectors of typhus in most of Europe. It was used extensively by the US Army in early 1944 to halt an epidemic of typhus in Naples, Italy (Beatty, 1973). Beatty also reported that DDT was effective for control of the mosquito *Aedes aegypti,* the vector of yellow fever in South America and Africa, and it killed the flea that vectored

---

4. http://14.139.56.90/bitstream/1/2033800/1/34.pdf. Accessed April 23, 2015.

5. www.appalachianhistory.net/2012/03/army-used-ddt-for-de-lousing.html. Accessed November 12, 2013.

6. Pyrethroids are synthetic insecticides based on natural pyrethrum. Pyrethrins are insecticides derived from flowers of the *Chrysanthemum* genus, many of which are cultivated as ornamentals. There are more than 3500 registered insecticides based on pyrethrum.

bubonic plague. In the South Pacific, aerial spraying gave spectacular results for control of malaria and dengue fever. Malaria was, in fact, completely eradicated from many islands. DDT formulations also had great value in agricultural entomology and stimulated the search for effective insecticides from other chemical families.

It will surprise many to learn that "for his discovery of the high efficiency of DDT as a contact poison against several arthropods" Paul Hermann Müller was awarded the 1948 Nobel Prize in Physiology and Medicine. Müller's was the only Nobel that has ever recognized pesticide research. He was born at Olten, Solothurn, Switzerland, on January 12, 1899 (died October 12, 1965). The family moved to Basel where Paul attended primary, elementary, secondary school, and Basel University. He received his doctorate in 1925.

In his Nobel address, Müller said "We are still far removed from being able to predict with any degree of reliability, the physiological activity to be expected from any given constitution. In other words, connexion between constitution and action are so far still quite unexplained. In addition we have the particularly difficult conditions caused by the uncertainty of tests on living material. More difficult still are the relationships in the field of pesticides, and in particular of synthetic insecticides."... "There are no points of reference to begin with so that we can only proceed by feeling our way." Although great progress has been made in study of pesticide chemicals structure and activity relations, in the pesticide realm, scientists are still feeling their way.

In Müller's view, the properties of an ideal insecticide were: great insect toxicity, rapid onset of action, little or no mammalian or plant toxicity, no irritant effect, a wide range of action, long persistence, and low price. At the beginning of World War II, DDT satisfied all of his criteria.

Its first widespread use as a pesticide against human disease was during World War II. It was so effective that it was called the atomic bomb of pesticides. Müller and the Geigy Corporation and several other companies that began to produce and sell DDT during World War II gave the world something new. It gave better insect control than any of its predecessors, was very effective against agricultural insect pests, and less toxic than other available agricultural insecticides. It had the ideal properties that Müller described. Similarly, the American Chemical Paint Company (see Chapter 6) and several following companies gave the world 2,4-D after World War II. It gave better, selective, lower cost, and safer weed control than its predecessors. Both received praise for their contributions because they represented what Kinkela (2011, p. 107) called the "curative power of modern technologies." Only later were they, all other pesticides, users, and those in the pesticide business criticized for what they were doing that nearly all had previously regarded as good and necessary progress.

SYNTHESIS

DDT is a member of a large group of chlorocarbon compounds (chlorinated hydrocarbons), that includes the following:

1. Chemicals that keep us cool—what Le Couteur and Burreson (2003) call the fabulous freons. They are a variety of nonflammable gaseous or liquid fluorinated hydrocarbons employed in refrigeration, air conditioning, and as aerosol propellants.
2. PCBs—chemicals ideal for use as electrical insulators and coolants because they are stable at very high temperatures and essentially inflammable. They were also used as plasticizers in a wide range of products including paint, waxes, lubricants, adhesives, plastic wraps, coffee cups, and baby bottles. All are based on the biphenyl molecule (Figure 7.3).
3. Combustion by-products—dioxins and furans are chlorocarbon compounds in a family of toxic substances that share a similar chemical structure. Most are not produced intentionally. They are created when other chemicals or products are made. The most notorious is 2,3,7,8-tetrachloro-p-dibenzo-dioxin. It is a highly toxic teratogen (causes birth defects). Several dioxins exist but this particular one was created as an undesirable by-product when temperature was not properly controlled in the production of the 2,4,5-T for use in agent orange (see Chapter 6).
4. The last group of chlorocarbon compounds of concern is pesticides that contain chlorine. DDT and 2,4-D are prime examples. There are hundreds of others.

DDT is produced by the reaction of chloral (CCl_3CHO) with chlorobenzene (C_6H_5Cl) in the presence of sulfuric acid, which acts as a catalyst. DDT can also be regarded as a derivative of 1,1-diphenylethane (Figure 7.4). The commercial

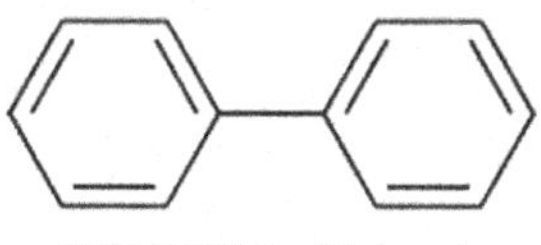

FIGURE 7.3 Biphenyl.

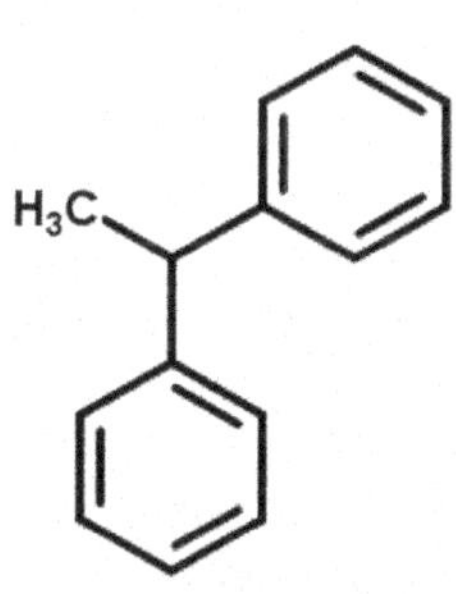

FIGURE 7.4 Diphenylethane.

formulations are a mixture of several closely related isomers. The major component (77%) is the *p,p'* isomer. The *o,p'* isomer is usually present in amounts around 15%.

DDT is highly hydrophobic—nearly insoluble in water—but quite soluble in most organic solvents, fats, and oils. It persists in the environment because it is readily adsorbed by soil clay colloids (e.g., montmorillonite). Soil is a sink and a source of long-term exposure. DDT's half-life in soil ranges from 22 to 30 days, but anaerobic and/or aerobic degradation occurs only after desorption from soil colloids. It can escape from soil and into the environment by runoff, volatilization, and photolysis. Its lipophilicity and environmental persistence were initially regarded as favorable traits that improved its efficacy. They are also the primary factors that led to its rapid descent into ignominy.

SILENT SPRING

More than 50 years ago, Rachel Carson, a marine biologist, published Silent Spring (1962), a polemic that exposed the dangers of chemical pesticides—including, most notably, the organochlorine (chlorocarbon) insecticide DDT. Silent Spring was first published in mid-1962 as a series of articles in the New Yorker. After the book was published in 1962, it became an influential bestseller. It was, without doubt, the primary impetus behind the successful movement to ban DDT, which the book, with a touch of hyperbole, characterized as deadly. Silent Spring also served as an important stimulus for the US environmental movement and creation of the US Environmental Protection Agency in December 1970. Kinkela (2011, p. 112) correctly points out a central point of Carson's book that was most likely not noted or, if noted, was conveniently neglected by her critics: Carson did not advocate elimination of all pesticides. She acknowledged that they had great potential if applied in proper doses with great care. She said:

> *It is not my contention that chemical insecticides must never*
> *be used. I do contend that we have put poisonous and biologically*
> *potent chemicals indiscriminately into the hands of persons*
> *largely or wholly ignorant of their potential for harm.*

She believed that application (spraying) of large quantities of any pesticide can have serious environmental consequences. She urged caution about synthetic chemicals—look before you leap and you may choose not to leap. She encouraged regular consideration and careful thought about potential harm to humans and other species with whom we share the planet. Her particular focus was the effects of DDT on birds. Spring would be silent when the birds were gone. Her view was that these new pesticides were like attractive, persuasive devils that, without telling us, could create change, poisoning, and death. The ensuing discussion which Carson began has not been silent.

Her book reflected the Enlightenment, a philosophical movement of the seventeenth and eighteenth centuries characterized by belief in the power of human reason. It was concerned with a critical examination of previously accepted doctrines from the point of view of rationalism, which argued that reason is the only valid basis for action. Reason is a way, arguably the best way, to make sense of things. It establishes and verifies facts that justify practices, beliefs, conclusions, and judgments.

Empirical observation and scientific authority asserted that pesticides were a vast technological improvement and DDT was one of the most outstanding. Carson's polemic, her reasons, against pesticides combined with scientist's confidence in their contributions to public health and agricultural productivity helped people begin to believe that science might be doing more harm than good. Carson argues that "civilisation was uncivilised and economic growth was destroying, not creating, the things in life that were of real value" (Economist, 2013, p. 7).

The Enlightenment, as Silent Spring did two centuries later, changed attitudes about the rights of others. In the seventeenth century, people were not expected to consider the well-being of anyone beyond their family or tribe. Enlightenment scholars broadened the scope of moral responsibility to include compatriots and, later, foreigners (Economist, 2013, p. 6). Now Carson's assertion that moral responsibility should be extended to other creatures that have rights we should respect and protect has been widely accepted. In addition, the environment is now regarded as deserving of moral consideration.

Environmental concern and debate were undoubtedly strengthened by the first Earth Day in 1970. Environmentalists were commonly regarded as troublesome if not simply "wackos" by agricultural people and the agrichemical industry. Rising environmental awareness and concern stimulated philosophers to develop a philosophical foundation for concern about the environment. The serendipitous confluence of Earth Day and the presence and growing influence of environmental organizations created an intellectual climate that was stimulated by White's (1967) *The Historical Roots of our Ecological Crisis* and Hardin's (1968) *The Tragedy of the Commons*. Most influential was Aldo Leopold's (1968) A Sand County Almanac, in which he explicitly claimed that the roots of the ecological crisis were philosophical. Philosophers agreed.

The first academic conference of philosophers was organized by William Blackstone (1975) at the University of Georgia in 1972 and published as Philosophy and Environmental Crisis. White's and Hardin's were seminal papers. Cobb's book (1972) Is It Too Late? A Theology of Ecology was the first single-authored, environmental book by a philosopher. In 1975, environmental ethics came to the attention of mainstream philosophy with the publication of the Rolston (1975) paper: Is There an Ecological Ethic?[7] Rolston thereby created the philosophical foundation of environmental ethics.

---

7. My source for this history is http://www.cep.unt.edu/novice.html. For a longer description of environmental concern, see http://plato.stanford.edu/entries/ethics-environmental.

Subsequently, several authors provided substantial justification for why animals deserve moral consideration: Cavalieri (2001), Midgley (1984), Regan (1987, 2004), Regan and Singer (1989), Rollin (1989, 1995, 2006, 2011), Singer (1981, 1985, 2002, 2004), Zimdahl (2012), and others. There has been change, albeit a slow change, in our moral attitudes and perception of our moral duties to the natural world. Rachel Carson deserves much credit for her contribution to the expansion of our moral responsibility.

Unfortunately, many thought and some still think that Carson pushed the pendulum past the point of scientifically justified equilibrium and on to the side of costly paranoia about potentially live-saving pesticides. The history of state and US government hearings, organizational roles, scientific participation, and industrial reaction is available in Kinkela (2005). Subsequent research has shown that DDT is not as harmful to humans or birds as had been assumed, when the dose is appropriate to the task. It is harmful to mosquitos that transmit malaria and typhus that kill millions every year across the developing world.

It is impossible to know, but reasonable to suggest, that many human lives have been lost—particularly in tropical countries—as a result of DDT's vastly diminished use worldwide following publication of Carson's book and DDT's controversial ban by Ruckelhaus, Director of the US Environmental Protection Agency in 1972. The controversy was most apparent in the frequently strenuous, purportedly scientific, but actually nonscientific, thoughtful, yet undeniably emotional objections raised by employees of several companies of the growing agrochemical industry (e.g., American Cyanamid, Ciba, Dow, Geigy, Monsanto, Velsicol (see Kinkela, 2011)). Their case was supported by Congressman Whitten's (1966), equally polemic, presentation of "the facts about pesticides." The industry's claim was that "the DDT conflict set a pattern for environmental decisions based less on scientific evaluation of risk than on histrionics, mudslinging, and hyperbole" (Kelly, 1981). Kinkela (2011, p. 120) correctly points out that the chemical industry's arguments did not just question Carson's legitimacy as a scientist; she was, after all, a woman. Their arguments "directly challenged the scientific legitimacy of ecology." Ecology and ecologists' comments were dismissed by the pesticide community and by most agricultural scientists. Ecologists were too emotional, not scientific, and wrong. Historical study has clearly demonstrated that both sides of the intense debate were not at all averse to histrionics and hyperbole. Beatty (1973) objected to the "immense accumulation of inaccurate information and emotionalism about pesticides, especially DDT." In her view, it placed the United States and the world in an untenable position regarding insect control in agriculture and protection of human health. Beatty suggested the environmentalists (at the time, an epithet) and the amateurs were winning.

ENVIRONMENTAL EFFECTS

Rachel Carson and many following writers are credited with advancing the global environmental movement. However, one must ask was Carson's concern appropriate? Was she right? The answer, without question, is yes. The primary

TABLE 7.1 Audubon Society Data from Annual Christmas Bird Counts

Year	Number of Birds Counted in the United States
1950	9,321,081
1960	6,862,842
1970	87,275,598
1980	106,305,151
1990	116,633,849[a]
2000	75,279,544
2010	52,457,750
Average since inception	65–70 million/year

[a]It is tempting, but incorrect, to interpret these large numbers as a sudden increase in the bird population. The Audubon Society attributes these counts to one or all of three things: an increase in the number of people counting, an increased number of places where counts occur, and the strong influence of a huge count of one or two species in a few places. Personal communication, January 17, 2014, Mr Geoff LeBaron, the National Audubon Society, New York.

focus of her concern was the effect of DDT on birds, especially eagles, ospreys, and other large raptors, that unavoidably crushed their own eggs. However, the annual Audubon Society Christmas bird count of all species was more than five times greater in 2010 than in 1950, when DDT was introduced (Table 7.1). The Audubon Christmas bird count is affected by the time of year (birds not present in the winter cannot, of course, be counted), the number of observers, and that the 1980 and 1990 counts were affected by a huge population of a single species, primarily in southeastern United States. The Breeding Bird survey conducted by the US Fish and Wildlife Service of the Department of the Interior is conducted in May or June at more than 2000 locations across the United States and southern Canada. It monitors changes in breeding birds across the continent and allows close monitoring of the rate of change in areas where rapid changes are occurring. It is an early warning system. Both studies confirm that Carson's concern was appropriate. Both studies affirm that agricultural practice is one of the important, but not the only factor, that affects bird populations.

The American robin population, which Carson noted in her discussion of DDT's effects, has had a strong recovery across the continent (Robbins et al., 1986). This evidence does not, of course, absolve DDT of the scientifically documented effects on songbirds and eggshells. In addition, Audubon data showed that the population of 20 common birds had declined an average of 68% over the past several years.[8] They do not attribute the decline directly to DDT or any other pesticide.

---

8. Personal communication. December 2013. Ms Kathy Dale. Director of Citizen Science, National Audubon Society, Washington, DC.

They cite the combination of suburban sprawl, industrial development, agricultural intensification, and pesticides as the likely causes. Agricultural intensification undeniably includes increased use of pesticides. Freemark and Kirk (2001) note that the "population decline of farmland birds over recent decades in Europe, Canada, and the US has been attributed to more intensive agricultural management." They go on to emphasize the importance of noncrop habitats, more permanent crop cover, and less intensive management practices to the conservation birds on farmland. Mineau and Whiteside (2013) agree with Freemark and Kirk that agricultural intensification is important, but for some species the indirect effect of insecticides and fungicides on bird food sources is an important factor in bird decline.

At the same time, one must ask, was DDT all bad? Is it true that it had no beneficial effects? There is no question that DDT was and still is an effective insecticide. Its efficacy is determined by the same characteristics that made it an environmental problem: persistence, lipid solubility (lipophilicity), and high toxicity to a variety of insects. In other words, it was not selective. It killed many different kinds of insects, some harmful and some beneficial, but high toxicity to insects was combined with low toxicity to other organisms, including humans. It was also readily accepted because of its low cost. One of the environmental problems with DDT occurred because of its success. It worked so well that many people assumed that if the recommended amount worked well more would be even better. "Enthusiastic use became enthusiastic overuse" (Kelly, 1981). With DDT, other pesticides, and other environmental chemicals, environmental pollution and subsequent harm to other creatures were inevitable. It soon became evident that DDT was everywhere. Therefore, when problems occurred, DDT was blamed.

By 1970, 3 million tons of DDT had been manufactured and used, dominantly in agriculture. More than 40,000 tons were used in the world each subsequent year. In the United States, it was manufactured by about 15 companies whose production peaked in 1963 at 82,000 tons. About 1.4 billion pounds were applied in the United States before the 1972 ban, with peak usage of 36,000 tons in1959.[9] After 1959, DDT use in the United States declined greatly, dropping to just under 12 million pounds in the early 1970s, before it was banned in 1972.

Scientists have claimed that DDT is so stable that it may never be completely eliminated from the environment. DDT can still be found at very low concentrations in soil and water and, although there is great concern about the long-term effects of its presence, there is no scientific evidence of harm to humans from the existing low concentrations.

HUMAN EFFECTS

There is no firm scientific evidence that DDT causes cancer or is directly harmful to humans. There have been no reported human deaths from DDT. It

---

9. HTTP://en.wikipedia.org/wiki/DDT. Accessed October 2013.

is a chlorinated insecticide that kills insects by interfering with nerve control processes unique to insects. When used for insect control at doses prescribed by scientific research, DDT is not acutely toxic to mammals. The LD_{50} (the lethal dose at which 50% of a test population (usually mice or rats) dies) is 113–118 mg/kg in rats and 150–300 in mice. The accepted fatal dose for humans, after oral consumption, is approximately 30 g = 30,000 mg.

It does persist and there is still evidence that it can be found in human mother's milk (Mannetje et al., 2014). That, although certainly unacceptable, has not been shown to be harmful. Its oral LD_{50} in rats is 87 mg/kg (ppm),[10] making it moderately hazardous by World Health Organization standards. It was used extensively during World War II as a delousing powder to successfully stop the spread of typhus and kill the larvae of disease carrying mosquitoes. Bug bombs, made from aerosol cans filled with DDT, were used by the US military in the South Pacific. They controlled malaria while releasing large amounts of chlorofluorocarbons, the propellant, and DDT into the environment. Soldiers and the military commanders didn't worry about either. They worried about malaria (Le Couteur and Burreson, 2003).

In spite of the obvious agricultural and public health benefits of DDT, after Rachel Carson, many argued strongly that human health and environmental risks justified its being banned. Without question, pesticides of all kinds are meant to kill living organisms (pests); therefore, it is not illogical to assume that they must be harmful to humans. Several human health concerns have been reported about DDT or its primary degradation product DDE. These include endocrine disruption, premature birth and low birth weight, developmental neurotoxicity from in utero exposure, decrease in semen quality, possible association with miscarriage, and possible role in incidence and/or causation of cretinism (specific references can be found in http://en.wikipedia.org/wiki/DDT). Replicated scientific evidence, which supports claims of adverse human health effects from DDT exposure, is rare to nonexistent (Attaran, 2001). Attaran and Maharaj (2000) argued that DDT should not be banned because of its clear public health benefits. Subsequent to the US ban, the 2001 United Nations Convention on Persistent Organic Pollutants drafted an international agreement to limit DDT's use and prohibit the use of other persistent pesticides (Kinkela, 2011, p. 4). The verified environmental problems cited included that the pesticides were toxic, resisted degradation in soil, bioaccumulated, were transported throughout the environment in air and water across international boundaries, and they had negative effects on several species. Kinkela (2005) argued that the DDT conflict situated "environmental ethics squarely at odds with the ethics of saving lives with a known environmental pollutant."

---

10. For comparison: the oral (in rats) LD_{50} in ppm = mg/kg of a few other chemical compounds. Aspirin = 250, caffeine = 192, DDT = 87, ethyl alcohol = 7060, unleaded gasoline = 4, pyrethrins = 200, table salt = 3000, 2,4-D = 375–666.

MALARIA

In 1955, the World Health Organization began a program to eradicate malaria from the world. It was based on widespread use of DDT. Initially it was highly successful and eliminated the disease in most of the Caribbean, the Balkans, some of North Africa, northern Australia, Taiwan, and many South Pacific islands. Malaria is the most significant human parasitic disease. Worldwide, there were an estimated 207 million cases and 627,000 deaths in 2012. Annually it is the infectious disease that is a major cause of early death of children.

In India, after 10 years of DDT use, malaria deaths declined from 750,000 to 1500/year. In what was then Ceylon (now Sri Lanka), DDT reduced the number of malaria cases from 2.8 million in 1946 to 110 in 1961. Seven years after DDT use stopped in Sri Lanka, the number of cases was 2.5 million (Kelly, 1981). There was no other inexpensive, effective, easy to use insecticide that during use had no obvious harmful effects on humans and was as effective against malaria transmitting mosquitoes. South Africa was malaria free after it began using DDT in the 1940s. Effective pressure from environmental groups compelled a switch to other insecticides in 1996. A particularly aggressive mosquito species (*Anopheles funestus*)[11] invaded. After a 15-year absence, malaria cases rose from 4117 in 1995 to 27,238 in 1999 and possibly many more. Swaziland never stopped using DDT, and there was no increase malaria cases. South Africa returned to DDT in 2000 (Attaran, 2001). These were major, perhaps unique, public health achievements.

Malaria came out of central Africa and traveled all over the world. It probably contributed to the fall of the Roman Empire. Malaria genes have been found in the bones of buried children in Roman ruins dating to 450 AD. The British invented gin and tonic after they colonized India and found endemic malaria. The tonic in gin and tonic—quinine—relieved malaria's symptoms, but did not cure it. Soldiers wouldn't drink quinine because it is bitter. Adding gin and a bit of lime to combat scurvy made quinine tolerable.

The name malaria comes from Italian *mal* (bad) and *aria* (air = bad air), because it was believed evil swamp gases caused it. Quinine was used to treat malaria in Rome in 1631. It was endemic in the swamps and marshes around Rome. Malaria killed several popes, many cardinals, and countless Romans, Oliver Cromwell, Dante, and Alexander the Great, Lord Byron, and eight American presidents (including President Kennedy) had repeated bouts with it.

When malaria attacks, the hemoglobin of the blood is converted to hemozoin or hematin, neither of which carries oxygen. The infection remains in the body for life. The long-term effect is destruction of red corpuscles and loss

---

11. Of the 3500 mosquito species in 41 genera, those that transmit malaria all belong to the *Anopheles* genus. Only 30–40 species of the *Anopheles* genus transmit *Plasmodium falciparum* the parasite that causes malaria. The male is innocent because the male proboscis is incapable of penetrating skin. (https://www.google.com/webhp?sourceid=chrome-instant&ion=1&espv=2&ie=UTF-8#q= world+deaths+from+malaria+and+other+infectious+diseases. Accessed November 2014.

of hemoglobin. Anemia and melanosis (darkening of the skin) are the visible effects. The spleen enlarges and the liver is damaged.

From colonial times until the 1940s, malaria was the American disease. It thrived from New York to Florida and from North Carolina to California. Up to 7 million Americans were affected annually until the mid-1920s. In 1936, 3900 Americans died from malaria. One of the Continental Congress' first expenditures was $300 worth of quinine to protect Washington and his troops from malaria. By the 1950s, malaria was well on its way to being eradicated in the developed world. DDT and mosquito eradication programs solved, but have not eliminated malaria in the United States and most developed countries. Unfortunately, but predictably, widespread agricultural use of DDT led to insect resistance and, in spite of the success of early malaria eradication campaigns, a resurgence of malaria followed.

DDT kills *Anopheles* mosquitoes, and *Anopheles gambiae*, the primary carrier of the protozoan parasites that cause malaria is the common target. In humans, *Plasmodium falciparum* causes 75% of the cases and the majority of deaths, followed by *Plasmodium vivax* (20%) and *Plasmodium malariae*, *Plasmodium ovale*, and *Plasmodium knowlesi*. The World Health Organization estimates that in 2010 there were 219 million cases of malaria resulting in 660,000 deaths (World Malaria Report, 2012). Olupot-Oulpot and Maitland (2013) estimated the number of cases between 350 and 550 million for *P. falciparum* malaria. Lozano et al. (2012), in an exhaustive study of worldwide deaths from diseases of all kinds, estimated 975 million deaths from malaria in 1990 and 1.17 million in 2010, an increase of 20%. Murray et al. (2012) estimated 1.24 million deaths in 2010, up from 1 million in 1990. The majority of cases (65%) occur in children younger than age 15 years (Murray et al.).

At present, malaria can be prevented and treated, but not cured. There is no vaccine. Protection against infection is mosquito-focused: protective clothing, repellents, bed nets, and mosquito control. Studies in Ghana, Kenya, and other African nations show that about 30% of child deaths could be prevented if children slept under bed nets regularly treated with insecticides. In spite of the fact that DDT made global eradication seem possible, it was banned or abandoned because of its well-known environmental and ecological effects. It is ironic that it is the insecticide of choice to treat bed nets, a primary defense. Endemic malaria (e.g., sub-Saharan Africa) often increases birth rate because if one is sure some children will die, it is logical to have more.

It might be that Germany would have been much less successful in two world wars without its synthetic chemical industry, which began with a search for quinine, the drug that did not cure, but relieved malaria's symptoms. Quinine, like tea, was a product the Americans were willing to leave to foreigners, until the Japanese began to threaten in the 1930s and the quinine supply was jeopardized. The high cost of the bark of the quinine tree (*Cinchona officinalis* L.), its low efficacy (purity), and limited supply led organic chemists to coal-tar chemistry. The quinine search began in the 1830s.

Le Couteur and Burreson (2003) report that William Henry Perkin, a student at London's Royal College of Chemistry, decided to try to synthesize quinine. The chemical formula ($C_{20}H_{24}N_2O_2$) had been determined, but the structure was unknown. Perkin and several of his chemical colleagues were "convinced that quinine could be synthesized from materials found in coal tar." His experiments, conducted in a laboratory in his home, were not successful. But as is true with so many experiments, although he failed in his primary objective, he achieved "the first true multi-step synthesis of an organic compound." His work led to the creation of colored dyes from coal tar residue. Perkins unsuccessful experiments to synthesize quinine, created the chemical dye industry. He retired a wealthy man.

Quinine's chemical structure was determined in 1918 by the German chemists Rabe and Kindler (1918). It was not until 1944 that the American chemists Woodward (Nobel laureate 1965) and Doering (1944) synthesized quinine. German chemists were the best in the world in the nineteenth and early twentieth centuries. In addition to their work on quinine and Perkins' creation of dyestuffs, German chemists created explosives 1860s, saccharin 1892, aspirin 1904, viscose-rayon and cellulose 1910s, synthetic rubber and petroleum 1920s, and sulfa drugs in the 1930s. One must wonder what the world would be like if the British or the Americans had developed all the products of organic chemistry Germans chemists did. In explosives, dyes, fertilizers, pharmaceuticals, and synthetic petroleum, German chemists enabled Weimar Germany to lead the world in chemistry and into war.

AGRICULTURAL EFFECTS

Shaw (1946) described the early agricultural view of the chlorinated hydrocarbons, a new class of insecticides that offered previously impossible insect control. DDT was the most widely used chlorinated hydrocarbon. It transformed agriculture because it offered environmental stability and persistence and thus insect control over time that was vastly superior to control achieved by the available, mostly contact insecticides. It was not just another step in the development of methods of insect control. It was a revolutionary, order of magnitude change.

Entomology, the oldest pest control discipline, has a 200+ year history. It was a descriptive science long before it became quantitative and theoretical. It was begun by people interested in insects. Many of the founders were amateurs with little and often no formal education who were fascinated by insects. In the mid-eighteenth century, entomologists began to systematize their science by study of the taxonomy and the life history of insects (Smith et al., 1973, p. 96). Then and well into the nineteenth century, entomologists had no good techniques to prevent damage to crop yield or to protect people from insect-borne diseases. Insects could be observed and described, but they could not be controlled; their populations could not be managed. The discipline changed rapidly after World War II when DDT and other synthetic organo-chlorine insecticides gave entomologists the ability to selectively control insects.

The primary purpose of all agricultural insecticides is prevention of insect damage and thus protection of yield. However, it is important to recognize that the advent of DDT and other insecticides did not increase crop yield, they protected it. In the world of agricultural and public health insect control, DDT was good news. Shaw knew a great deal of experimental work had been done but not published because of World War II. In his view, lack of publication permitted wild speculation and exaggeration rather than what he assumed would be a calm, restrained, balanced scientific approach. What followed, in a very real sense, what DDT created, was extensive, widespread scientific research and vigorous dispute and disagreement.

In 1946, Shaw reported that DDT gave better control than available insecticides for some 30 agriculturally important insects, was only about equal against 19 and inferior against 14. It was an improvement, not a panacea. But it was a huge change. It was the first of many subsequent insecticides that gave farmers and entomologists the ability to control a variety of insects. As entomologists developed and perfected insect management strategies, they also recognized, studied, and were concerned about the risk of harm to beneficial insects, birds, fish, and larger animals, especially after regular use over large areas. Most scientists agreed that because DDT showed such promise that it would be unfortunate if its future use were prejudiced by early misuse.

Kinkela (2011, Chapter 3) provides much greater detail on the development of the green revolution in Mexico than is warranted here. The green revolution was and will continue to be regarded as a major achievement of agricultural science. Some of the stimulus was the 1940 visit of Henry Wallace, then vice president of the United States, to Mexico and his observations of the poor state of its agriculture. He and President Roosevelt saw an opportunity for United States to be involved in the development of Mexico and perhaps all of Latin America. Subsequently officials of the Rockefeller and Ford foundations became involved. Several forward-looking thinkers were confident the United States would win World War II. They asked and discussed what the world's major problems would be when the war ended. They decided that the most important problem would be producing enough food to feed the poor and those whose agricultural resource—land and infrastructure—had been devastated by the war. They proceeded to provide funds to establish the international center for wheat and corn CIMMYT (Centro Internacional de Mejoramiento de Maíz y Trigo) in Mexico in 1943 followed by the International Rice Research Institute (IRRI) in the Philippines in 1959. The scientists at these two international institutions were, in a very real sense, the creators of the green revolution. CIMMYT and IRRI are now part of the Consultative Group on International Agricultural Research. It includes 15 independent, nonprofit international agricultural research centers[12] whose primary mission is to continue improvement of the productivity and environmental sustainability of world agriculture.

What did the green revolution accomplish? This very long story will be abbreviated to emphasize the role of DDT and other pesticides. The first enduring accomplishment of the green revolution was prevention of starvation of millions of people after World War II. The goal of feeding all an improved diet continues to be primary among all international agricultural research institutes. To achieve this worthy goal a wide array of dedicated agricultural scientists combine their talents to create a new way to practice agriculture—a new farming system—that includes carefully selected new crop varieties, appropriate fertilizers, irrigation where required, and pesticides for insect and weed control. Much of this is accomplished by carefully selecting new, improved crop varieties with the genetic potential for higher yield. Crop varieties are also selected for improved response to fertilization, resistance to lodging, shorter time to maturity, and reduced sensitivity to day length. The green revolution and subsequent international and national research accomplishments were and continue to be superb scientific achievements, which because they are government- and foundation-supported, are distributed without charge to other research institutions, to the public, and to farmers. Scientists accomplished what they set out to do—yields increased and massive starvation was prevented. The green revolution was also a social phenomenon because of its effects on agricultural development, self-confidence among farmers and the spread of new technologies.

The technologies of the green revolution included the pesticides to achieve and maintain production. Pesticides including DDT became part, indeed an essential part, of a new way to practice agriculture. Their necessity was justified by their contribution to increased yields that characterized the green revolution. The new agricultural system was energy, capital, and chemically intensive. It was, in the view of many, the transfer of American technology with all of its presumed advantages and, unfortunately, its disadvantages to agricultural systems that were ill-adapted economically and technologically to take full advantage of the technological array. Proponents of the green revolution claimed that the alternative to the new agricultural system was famine.

The ready availability and advantages of DDT and other organochlorine insecticides plus an array of other pesticides changed what one might call the chemical landscape of American agriculture. Several chemical companies began synthesis and testing candidate pesticides and promoted their use in

---

12. In addition to CIMMYT and IRRI, the other centers that are: Africa Rice Center, Benin; Bioversity International, Italy; Center for International Forestry Research, Indonesia; International Center for Agricultural Research in the Dry Areas, Lebanon; International Center for Tropical Agriculture, Colombia; International Crops Research Institute for the Semi-Arid Tropics, India; International Food Policy Research Institute, United States; International Institute of Tropical Agriculture, Nigeria; International Livestock Research Institute, Kenya; International Potato Center, Peru; International Water Management Institute, Sri Lanka; World Agroforestry Center, Kenya; and World Fish, Malaysia.

agriculture. Their advantages: selectivity, persistence, assumed safety to the user, low-cost, and effectiveness for pest control made them attractive and heralded a new age for American and developed country agriculture. Because the green revolution is dependent upon and encouraged transfer of American technology to the world, there was a transformation of agriculture in lesser developed countries.

More than 80% of the DDT used in the United States in 1970–1972 was applied to cotton to control the boll weevil. The rest was used on peanuts and soybeans. The decline in DDT's agricultural use was the result of a combination of the development of predictable (but largely ignored) insect resistance, more effective alternative pesticides (primarily the highly toxic organophosphates), growing public concern, with increasing scientific support, about the adverse environmental effects, and increasing government restrictions on use.

It soon became obvious that a second major effect of the advent of DDT for agricultural insect control and malaria prevention was that it soon became the bane, but not the death, of the growing agricultural pesticide industry that, in a very real sense, had begun because of the efficacy and presumed safety of DDT.

The third major effect of DDT is related to the fact that it was the first and remains the primary example of a good and bad pesticide. Improved, effective management of agricultural insect pest problems and the undeniable public health benefits have been clear. The equally undeniable environmental, avian and other pollution problems were first brought to the public's attention by Rachel Carson and the successes and failures of DDT.

DDT's undeniable benefit of improved insect control, its contribution to the development of pesticide chemical industry, and its real and imagined harm to the environment, humans and other creatures transformed American agriculture. The public and scientific demand for answers to the questions that, before DDT, had not been asked is its legacy. Its relatives, and subsequent pesticides from many chemical families contributed to the transformation of agriculture from what was regarded as a beneficial enterprise that produced the food we need to a suspect enterprise that has externalized[13] many of its costs. Agriculture and its practitioners had always been respected, but suddenly they were suspect, if not automatically guilty, because the public feared that in addition to food, agricultural practice produced undesirable, indeed unacceptable, human and environmental disbenefits.

When public concern about DDT and other pesticides was combined with the increasing role of environmental organizations (e.g., the Environmental Defense Fund, Friends of the Earth, Greenpeace, The World Wildlife Fund For Nature), and a firm philosophical foundation the public supported, if not demanded, agricultural consideration of and answers to new difficult moral and scientific questions.

---

13. An externality is a cost or benefit which affects a party that did not choose to incur that cost or benefit. For example, in agriculture if a pesticide harms humans, other creatures or the environment, neither the user, the seller, nor the manufacturer pays the cost, although they receive all the benefit.

1. How important is control of agricultural pests when a result is environmental and human contamination with potentially unknown short- and long-term effects.
2. Is our food safe?
3. Are pesticides necessary or are they a new technology that improves production efficiency, provide a significant profit to manufacturers, but cause irreparable harm to humans, other creatures, and the environment?

That demand and the ensuing discussion and debate are still going on. Agricultural practitioners and the supporting scientific enterprise initially resisted, but are now facing and dealing with the environmental and ethical questions that began with DDT.

REFERENCES

Anon, 2013. All creatures great and small: special report – Biodiversity. Economist September 14, 16.

Attaran, A., 2001. In Praise of DDT. Pesticide Outlook. June. p.83. Royal Society of Chemistry, London, UK.

Attaran, A., Maharaj, R., 2000. Doctoring malaria, badly: the global campaign to ban DDT. BMJ (formerly British Medical Journal) 321, 1403–1405.

Beatty, R.G., 1973. The DDT Myth: Triumph of the Amateurs. The John Day Company, New York. 201 pp.

Blackstone, W.T., 1975. Philosophy and Environmental Crisis. University of Georgia Press. 148 pp.

Carson, R., 1962. Silent Spring. Houghton Mifflin, Boston, MA. 368 pp.

Cavalieri, P., 2001. The Animal Question: Why Nonhuman Animals Deserve Human Rights. Oxford University Press, Oxford, UK. 184 pp.

Cobb Jr., J.B., 1972. Is it Too Late?: A Theology of Ecology. Bruce, Beverly Hills, CA. 147 pp.

Freemark, K.E., Kirk, D.A., 2001. Birds on organic and conventional farms in Ontario: partitioning effects of habitat and practices on species composition and abundance. Biological Conservation 101, 337–350.

Graham, J.R., 1921. Tar-heel War Record (In the Great World War – 1917–1918). World War Publishing Co., Charlotte, NC. 224 pp.

Hardin, G., 1968. The tragedy of the commons. Science 162 (3859), 1243–1248.

Kelly, H.E., 1981. The DDT debate: the beginning of the big ban era. ACSH News and Views 2 (3) 1, 13, 14. American Council on Science and Health, New York.

Kinkela, D., 2011. DDT and the American Century: Global Health, Environmental Politics, and the Pesticide that Changed the World. The University of North Carolina Press, Chapel Hill, NC. 256 pp.

Kinkela, D., 2005. The question of success and environmental ethics: revisiting the DDT controversy from a transnational perspective, 1967–1972. Ethics Place and Environment 8 (2), 159–179.

Le Couteur, P., Burreson, J., 2003. Napoleon's Buttons: 17 Molecules that Changed History. Jeremy P. Tarcher/Penguin, New York. 375 pp. Industrial and Engineering Chemistry Research 1994, 33, 2757–2763 2767.

Leopold, A., 1968. A Sand County Almanac, and Sketches Here and There. Oxford University Press, London; New York. 226 pp.

Lozano, R., 200 co-authors, 2012. Global and regional mortality from 235 causes of death for 20 age groups in 1990 and 2010: a systematic analysis for the Global Burden of Disease Study 2010. Lancet 380 (9859), 2095–2128.

Mannetje, A., Coakley, J., Bridgen, P., Smith, A.H., Read, D., Pearce, N., Douwes, J., 2014. Estimated intake of persistent organic pollutants through breast milk in New Zealand. New Zealand Medical Journal 127 (1401), 56–70.

Midgley, M., 1984. Animals and Why They Matter. University of Georgia Press, Athens, GA. 158 pp.

Mineau, P., Whiteside, M., 2013. Pesticide acute toxicity is a better correlate of U.S. grassland bird declines that agricultural intensification. PLoS One 8 (2), 1–8.

Murray, C.J., Rosenfeld, L.C., Lim, S.S., Andrews, K.G., Foreman, K.J., Haring, D., Fullman, N., Naghavi, M., Lozano, R., Lopez, A.D., 2012. Global malaria mortality between 1980 and 2010: a systematic analysis. Lancet 379 (9814), 413–431.

Olupot-Olupot, P., Maitland, K., 2013. Management of severe malaria: results from recent trials. Advances in Experimental Medicine and Biology 764, 241–250.

Rabe, P., Kindler, K.P., 1918. Berichte der Deutschen Chemischen Gesellschaft Uber die partielle Synthese des Chinins. Zur Kenntnis der China-Alkaloide XIX 51, 466–467.

Regan, T., 2004. The Case for Animal Rights. University of California Press, Berkeley, CA. 425 pp.

Regan, T., 1987. The Struggle for Animal Rights. Clarks, Summitt, PA. 197 pp.

Regan, T., Singer, P. (Eds.), 1989. Animal Rights and Human Obligations, second ed. Prentice-Hall, Englewood Cliffs, NJ. 280 pp.

Robbins, C.S., Bystrak, D., Geissler, P.H., 1986. The Breeding Bird Survey: Its First Fifteen Years, 1956–1979. U.S. Dept of the Interior, Fish and Wildlife Service, Washington, DC. Resource Pub. 157.

Rollin, B.E., 2006. Animal Rights and Human Morality. Prometheus Books, Amherst, NY. 400 pp.

Rollin, B.E., 2011. Putting the Horse before Descartes - My Life's Work on Behalf of Animals. Temple University Press, Philadelphia, PA. 285 pp.

Rollin, B.E., 1989. The Unheeded Cry – Animal Conscious Animal Pain, and Science. Oxford University Press, Oxford, UK. 330 pp.

Rollin, B.E., 1995. Farm Animal Welfare – Social, Bioethical, and Research Issues. Iowa State University Press, Ames, IA. 168 pp.

Rolston III, H., 1975. Is there an ecological ethic? Ethics 18 (2), 93–109.

Shaw, H., 1946. Some uses of DDT in agriculture. Nature 157, 285–287.

Singer, P., 2002. Animal Liberation. Harper Collins Publishers, New York. 324 pp.

Singer, P., 2004. Ethics beyond species and beyond instincts. In: Sunstein, C., Nussbaum, M. (Eds.), Animal Rights – Current Debates and New Directions. Oxford University Press, Oxford, UK, pp. 78–92. 338 pp.

Singer, P., 1985. In Defense of Animals. Harper and Row, New York, NY. 224 pp.

Singer, P., 1981. The Expanding Circle: Ethics, Evolution and Moral Progress. Princeton University Press, Princeton, NJ. 208 pp.

Smith, R.F., Mittler, T.E., Smith, C.N. (Eds.), 1973. History of Entomology. Annual Reviews, Inc, Palo Alto, CA. 517 pp.

White Jr., L., 1967. The historical roots of our ecological crisis. Science 155, 1203–1207.

Whitten, J.L., 1966. That We May Live: Here Are the Facts about the Effects of Pesticides on Our National Health—their Use, Dangers, and Contribution to Our Welfare—based on a Scientific Report Made in Behalf of Congress. D. Van Nostrand C. Inc., Princeton, NJ. 251 pp.

Woodward, R.B., Doering, W.E., 1944. The total synthesis of quinine. J. Am. Chem. Soc. 66 (5), 849.

World Malaria Report 2012. World Health Organization, Geneva, Switzerland. 195 pp.

Zimdahl, R.L., 2012. Agriculture's Ethical Horizon. Elsevier, London, UK. 274 pp.

Zoilkowski Jr., D., Pardieck, K., Sauer, J.R., 2010. On the road again for a bird survey that counts. Birding 41 (1), 32–40.

Chapter 8

Recombinant DNA

Chapter Outline

INTRODUCTION

The twentieth century included three outstanding scientific achievements that dramatically changed the way agriculture is practiced in the developed world. Each was essential to accomplishing agriculture's primary goal: feeding the world. The first, in 1905, was when Fritz Haber accomplished what chemists had long sought to do—fix nitrogen from air and combine it with hydrogen gas to synthesize ammonia. Carl Bosch, a chemist and engineer transformed Haber's tabletop method of fixing nitrogen into an important industrial process that eventually produced megatons of fertilizer: the Haber–Bosch process (see Chapter 4).

The second was Paul Müller's discovery of the high efficiency and insecticidal activity of dichlorodiphenyltrichloroethane (DDT). It is widely and simultaneously regarded as an environmental disaster and one of science's great public health achievements. It did not eliminate malaria but it saved many lives and relieved human suffering. DDT's undeniable benefit of improved insect control, its contribution to the development of pesticide chemical industry, and its real and imagined harm to the environment, humans, and other creatures transformed American agriculture (see Chapter 7).

The third was the rapid development of 2,4-dichlorophenoxyacetic acid (2,4-D) and the other phenoxyacetic acid herbicides for selective weed control in crops based on research during World War II, but not published until after the war. Selective weed control with synthetic organic chemicals transformed agricultural practice in much of the world and remains one of the great contributions of science that is also simultaneously deplored and applauded (see Chapter 6).

Haber, Bosch, and Müller became Nobel laureates. Each is properly lauded for the positive aspects of their scientific achievements. In each case, their scientific achievements and the subsequent positive transformation of agriculture had unexpected, negative consequences, as many scientific/technological developments have. Each of these scientific achievements became an essential part of a new way to practice agriculture. The Haber–Bosch process enabled vastly more efficient production of explosives. Haber's research also led to the creation of chemical weapons used during World War I. DDT provided excellent insect control and, many argue, saved more lives (500 million[1]), than all the pharmaceutical drugs, including antibiotics, and created significant environmental problems. 2,4-D revolutionized agricultural weed control, led to the loss of agricultural jobs, increases in farm size, and had immediate unrealized, negative environmental effects. These agricultural stories are similar to the results of the well-known work of Albert Einstein, who received the Nobel Prize in Physics in 1921, 3 years after Haber. Einstein's theories of relativity, gravity, mass, and energy revealed some of the secrets of the cosmos and paved the way for nuclear energy and the atomic bomb.

The story of great success and often unanticipated, negative achievements continues with the story of how recombinant DNA transformed agriculture. Biotech/genetically modified (GM)/genetically modified organism (GMO) crops, a product of recombinant DNA technology and careful scientific research have been the most rapidly adopted agricultural technology ever developed (James, 2013).

Before proceeding, let's be sure we know what DNA is and what recombinant means. The initials DNA (the acronym) stand for deoxyribonucleic acid. It is the genetic material in all cells. The chromosomes inside the nucleus of a cell are made of DNA, a fine, tightly coiled helix (a spiral) that may be a meter (a bit more than a yard) long. The DNA molecule contains the code that is the instruction for making all organisms from humans to corn. Each gene is a code for particular proteins; the essential components of living things.

The adjective recombinant means of or resulting from new combinations of genetic material. As a noun, it refers to a cell or organism whose genetic complement results from recombination. The cell or organism has genetic material produced when segments of DNA from different sources are joined to produce recombinant DNA.

---

1. http://www.accesstoenergy.com/2012/01/29/ddt-and-malaria/. http://www.discoverthenetworks.org/viewSubCategory.asp?id=1259. Accessed June 2014.

Nucleic acids were discovered by Friedrich Miescher in 1869 (Dahm, 2008). They carry a cell's genetic information and are capable of self-replication and synthesis of RNA (ribonucleic acid). In the late nineteenth century, a German biochemist, Albrecht Kossel, discovered that nucleic acids were long-chain polymers of nucleotides made of sugar, phosphoric acid, and several nitrogen-containing bases for which he received the 1910 Nobel Prize in Physiology and Medicine. Nucleic acids are long-chain biological molecules. They are found in all living cells and viruses and, in the form of DNA and RNA, control cellular function and heredity.

They encode, transmit, and express genetic information through their sequence in the DNA and RNA structures.

In 1943, Oswald Avery (1877–1955) a Canadian-born American physician, a pioneer in immunochemistry, and one of the first molecular biologists, reported that DNA was the material of which genes and chromosomes are made (Avery et al., 2000). Avery proved that DNA carries genetic information and suggested DNA might actually be the gene. Most people at the time thought the genes would be made of proteins, not nucleic acids, but by the late 1940s, DNA was largely accepted as the genetic molecule. Scientists still needed to figure out the molecule's structure and understand how it worked.

Pauling et al. (1951) discovered that many proteins take the shape of an alpha helix, spiraled like a coiled spring, which forms the backbone of thousands of proteins. In 1950, biochemist Erwin Chargaff proposed that the arrangement of nitrogen bases in DNA varied widely, but the amount of certain bases always occurred in a 1:1 ratio. These discoveries were an important foundation for the later description of DNA's structure. It is two long chains of nucleotides twisted into a double helix and joined by hydrogen bonds between the complementary bases adenine (A), guanine (G) (both purines), cytosine (C), and thymine (T) (both pyrimidines). The sequence of nucleotide bases determines the proteins created and thus individual hereditary characteristics.

In the early 1950s, the race to discover the structure of DNA was on. At Cambridge University, Francis Crick (1916–2004) a molecular biologist, biophysicist, and neuroscientist, and James D. Watson, an American research fellow (born 1928) were impressed, especially by Pauling's work. Meanwhile, at King's College in London, Maurice Wilkins (1916–2004) and Rosalind Franklin (1920–1958) were also studying DNA. The Cambridge team's approach was to make physical models to narrow down structural possibilities and create an accurate picture of the molecule. The King's College team took an experimental approach, looking particularly at X-ray diffraction images of DNA (see Watson, 1968).

In 1951, Watson attended a lecture by Franklin, who had found that DNA can exist in two forms, depending on the relative humidity of the air. She deduced that the phosphate part of the molecule was on the outside. Watson (1968) returned to Cambridge with what he called "a vague recollection" of the facts Franklin had presented. Watson and Crick made an incorrect model, which caused the head of their unit to tell them to stop DNA research, but the subject kept coming up and they continued (Watson).

Franklin, working mostly alone, found that her X-ray diffractions showed that the wet/high-humidity form of DNA had all the characteristics of a helix, as Pauling et al. (1951) had predicted. She suspected that all DNA was helical but did not want to announce this until she had sufficient evidence of other possible forms. In early 1953, Crick showed Franklin's results to Watson, apparently without her knowledge or consent. Crick later admitted, "I'm afraid we always used to adopt—let's say, a patronizing attitude toward her.[2]" Watson and Crick took a crucial conceptual step, suggesting the molecule was made of two chains of nucleotides, each in a helix as Franklin had found, but one going up and the other going down. In 1952, Crick learned of Chargaff's et al. (1950, 1952) and Elson and Chargaff's (1952) research that showed the base pairs always occurred in a 1:1 ratio. He added that to the model, so that matching base pairs interlocked in the middle of the double helix to keep the distance between the chains constant.

Watson and Crick showed that each strand of the DNA molecule was a template for the other. During cell division, the two strands separate and a new half is built on each strand, just like the one before. DNA thus reproduces itself without changing its structure—except for occasional mutations. Their discovery enabled scientists to gain a better grasp of biology at its most basic level and expanded our knowledge of biological systems (Garrett, 2013, p. 48). New techniques and new technology developed that allowed scientists to understand how genes and DNA function, interact, and control biological mechanisms, indeed how they control life.

Because the structure fit the experimental data perfectly, it was almost immediately accepted. Discovery of DNA's structure is generally regarded as the most important biological/chemical work of the past 100 years. The research it enabled will be the major contributor to the agricultural scientific frontier for the next several decades. By 1962, when Watson, Crick, and Wilkins won the Nobel Prize for Physiology and Medicine "for their discoveries concerning the molecular structure of nucleic acids and its significance for information transfer in living material," Rosalind Franklin had died (in 1958, 38 years old). Only living people are nominated. The prize usually is awarded only to living recipients, although after nomination and acceptance, it has been awarded posthumously—E.A. Karlfeldt, Literature, 1931; D. Hammarskjöld, Peace, 1961; W. Vickrey, Economics, 1996; and R. Steinman, Physiology or Medicine, 2011. One wonders, if Franklin had been alive, would she have been nominated.

RNA is one of a family of large biological molecules that perform vital roles in the regulation and expression of genes.[3] The chemical structure of DNA and RNA are similar, with three significant differences.

---

2. http://www.itechpost.com/articles/5902/20130228/double-helix-discoverys-forgotten-hero-rosalind-franklin.htm. Accessed September 2014.
3. http://www.ncbe.reading.ac.uk/ncbe/gmfood/chymosin.html. Accessed August 2014.

1. RNA is a single ribose sugar strand.
2. RNA is less stable and more prone to hydrolysis because of the hydroxyl in the pentose ring.
3. In DNA, adenine is always paired with its complementary base thymine (A–T) and guanine is always paired with cytosine (G–C). The G–C pairing holds in RNA, but uracil (unmethylated thymine, a pyrimidine) replaces thymine, resulting in an A–U pairing; a small, but very significant difference. Messenger RNA carries information about protein sequence to the ribosomes where protein synthesis occurs.

Recombinant DNA molecules can be formed by laboratory methods of genetic recombination (e.g., molecular cloning) to bring together genetic material (DNA) from multiple sources, creating sequences that are not found in natural biological organisms. Recombinant DNA technology (i.e., genetic engineering (GE)) is possible because DNA molecules from all organisms share the same chemical structure. They differ only in the nucleotide sequence within an identical overall structure. DNA is the keeper of the all the information needed to create an organism (Washington, 2012). All DNA is made up of a base consisting of a sugar, phosphate, and one of four nitrogen bases, which, as mentioned previously, are always specifically paired (A–T and G–C). The sequence the nitrogen bases can be arranged in infinite ways in the double helix structure (Figure 8.1). The nitrogen

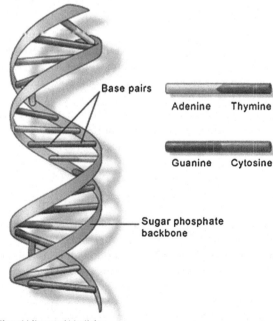

Base pairs

Adenine   Thymine

Guanine   Cytosine

Sugar phosphate backbone

U.S. National Library of Medicine

FIGURE 8.1   Structure of DNA. http://ghr.nlm.nih.gov/handbook/illustrations/dnastructure. *Accessed May 2014.*

bases are the same for all organisms. It is their sequence and number that creates diversity in the proteins and the resulting organisms.[4] It is beyond the scope or intent of this book to describe the process of creating a GM plant.

THE CASE FOR USE OF RECOMBINANT DNA IN AGRICULTURE

Recombinant DNA technology was made possible by the 1970 discovery, isolation, and application of restriction endonucleases (restriction enzymes) by Arber, Nathans, and Smith, for which they received the 1978 Nobel Prize in Medicine.[5] The idea of recombinant DNA was first proposed by Peter Lobban, a graduate student in Biochemistry at Stanford University Medical School. The first publications describing the successful production and intracellular replication of recombinant DNA appeared in 1972 and 1973. Stanford University applied for a US patent on recombinant DNA in 1974, which was awarded in 1980. The first licensed drug generated using recombinant DNA technology was human insulin, developed by Genentech and licensed by the Eli Lilly Corporation.[6]

The Flavr Savr Tomato

Recombinant DNA technology = GE = GM = transgenic crops, is now widely used in agriculture to create organisms (GM/GMOs) that produce GM crops. The first GM food was the Flavr Savr tomato, produced by the Calgene Company. It was the first commercially grown genetically engineered food to be granted a license for human consumption by the US Food and Drug Administration (FDA) in 1992. FDA concluded that the Flavr Savr was as safe for human consumption as tomatoes bred by conventional means. GM slowed tomato ripening, prevented softening, while preserving natural color and flavor. Flavr Savrs were first sold in 1994, but production stopped in 1997 because Calgene's development and marketing costs became excessive; the company was not profitable. It was acquired by the Monsanto Company in 1997. The acquisition was consistent with Monsanto's intent to bring biotechnology into agriculture and further up the food chain.

Chymosin

The Flavr Saver was the first GM crop, but cheese production was the first use of genetic modification in our food supply. Rennet, a complex of enzymes produced in any mammalian stomach, is often used especially in production of hard cheeses. It contains many enzymes, including a proteolytic enzyme that

---

4. http://www.rpi.edu/dept/chem-eng/Biotech-Environ/Projects00/rdna/rdna.html. Accessed May 2014.

5. http://en.wikipedia.org/wiki/Restriction_enzyme. Accessed May 2014.

6. http://en.wikipedia.org/wiki/Recombinant_DNA. Accessed May 2014.

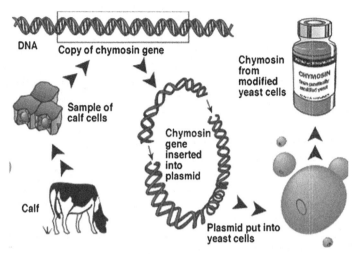

FIGURE 8.2   How the chymosin enzyme is made using genetically modified technology. *Source:* http://www.ncbe.reading.ac.uk/NCBE/GMFOOD/chymosin.html. *Accessed December 2014.*

coagulates milk, causing it to separate into solids (curds) and liquid (whey). The active (essential) enzyme for cheese production is chymosin. For years, the primary (perhaps the only) source was natural calf rennet, extracted from the inner mucosa of the fourth stomach (the abomasum) of young, unweaned calves. The stomachs were a by-product of veal production. Because of the limited availability of mammalian stomachs for rennet production, cheese makers had looked for other ways to coagulate milk since at least Roman times. There are many sources of enzymes, ranging from plants, fungi, and microbes that can substitute for animal rennet.

GE made it possible to extract rennet-producing genes from animal stomachs and insert them into certain bacteria, fungi, or yeasts to make them produce chymosin during fermentation. The GM microorganism is killed after fermentation and chymosin is isolated from the fermentation broth. Fermentation-produced chymosin has been used by cheese producers for more than 30 years. The cheese does not contain any GM component or ingredient. The chymosin is identical to that made by an animal, but is produced more efficiently. Chymosin is a product of genetic modification, but the cheese is not a GM product.

The American pharmaceutical company Pfizer perfected the GM technique in which RNA coding for chymosin is removed from the abomasum, inserted, using recombinant DNA technology via plasmids, into microbial DNA (e.g., the bacteria *Escherichia coli*) in a process known as gene splicing (Figure 8.2). The process, developed in 1981, received FDA approval in 1988. Through fermentation, the microbes possessing the bovine genetic material produce bovine chymosin, which is isolated and purified in quantities much

Percent of planted acres

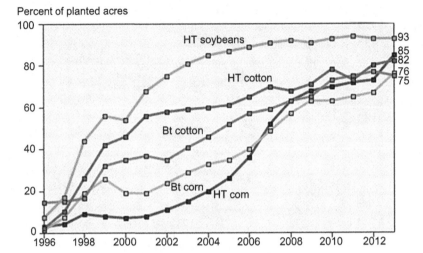

FIGURE 8.3 Adoption of genetically engineered crops in the United States, 1996–2013. http://www.ers.usda.gov/data-products/adoption-of-genetically-engineered-crops-in-the-us/recent-trends-in-ge-adoption.aspx#.U44BdvldUuc. *Accessed June 2014.*

greater than those that can be obtained from calf rennet and other sources (Figure 8.3).

Since GM chymosin in 1988 and Calgene's pioneering work on the Flavr Savr tomato in 1994, the number of GM crops has expanded rapidly. Recombinant DNA has become the dominant agricultural chemical technology. It is highly likely to become more important as agricultural land area is reduced, world population grows, and food demand (requirement) increases. Following are a few research areas where the promise of recombinant DNA technology may be fulfilled in agriculture and human health.[7]

Poppy to Yeast

GM/GE projects usually move a gene or genes from one species to another. This project moves genes from one creature to another: from opium poppy (*Papaver somniferum*) to yeast. The goal is to use yeast to manufacture beneficial, but addictive, opiates, such as morphine (chemically extracted from opium) and diacetylmorphine (heroin). Three crucial genes are transferred from opium poppy to yeast. It is good science. The justification is to cut opium poppies out of the loop thereby reducing, if not destroying, production and importation of these generally illegal drugs. An alternative, tightly controlled source would be created to produce necessary medical supplies (Economist, 2014).

---

7. http://www.ehow.com/info_8383532_uses-recombinant-dna-agriculture.html#ixzz32S3s6s7p. Accessed May 2014.

The Human Genome Project

Between 1990 and 2003, the United States funded and led the international Human Genome Project. It produced the genomic sequence of one human being. This has aided treatment of many diseases, identified disease-related genes, and developed more than 2000 diagnostic tests (Noble, 2013).

The developing field of synthetic biology, based on chemically synthesized DNA, combines several scientific and engineering disciplines to construct new biochemical production systems and organisms with new, specifically designed characteristics (Noble, 2013).

GM Technology in Agriculture

Since the first GM crops were introduced in 1996, the number of GM crops, other plants, and the number of scientists pursuing further uses of GM technology have expanded rapidly. The area planted and the number of crops have grown rapidly. The area planted (indicative of grower's acceptance) has grown[8] from 4.2 million acres in 1996 to 448.5 million acres in 2014 (ISAAA, 2015b); more than half in developing countries, more than 18 million of the world's farmers planted a GM crop in 2013 (James, 2013). Table 8.1 shows the GM area planted in 2013 of major crops in six major agricultural countries. The global agricultural biotech market was US$ 27.8 billion at the end of 2014; continued growth is predicted (ISAAA, 2015a). Table 8.2 shows some of the GM crops and other GM plants that have been approved for use. The total number of GM plants was 86 in 2014.[9] In 2011, 11 different GM crops were grown on 54% of the total crop area in developing countries. James (2013) asserts that "more than 90% (16.5 million) farmers that grew biotech crops in 2013 are risk-averse, small, resource-poor farmers in developing countries."

Many of these GM crops have been planted across the world (Table 8.3). Each is a significant part of the case for GM's continued, successful agricultural use.

It is worth noting that since 1998, no country has reported documented cases of harm to human health, development of allergies, or negative environmental effects from consumption of GM crops. It is also worth noting that, although the previous claim is true, there is no evidence that long-term, longitudinal studies have been conducted to look for evidence of harm (Thompson, 2007, p. 67). The reason such studies have not been done is, quite simply, they are very difficult to do. Conducting a longitudinal study of a pharmaceutical drug or a food additive enables isolation of the effect of the defined, often single object of study. That is not possible for a study of a whole food or the diet

---

8. http://www.gmo-compass.org/eng/agri_biotechnology/gmo_planting/142.countries_growing_gmos.html. Accessed September 2014.
9. http://www.gmo-compass.org/eng/database/plants/. Accessed September 2014.

TABLE 8.1 GM Area Planted and Crops in the Top 10 Countries in 2013 (James, 2013)

Country	Acres Planted (million)	Crops
United States	173.2 (68.5[a])	Corn (93), soybean (94), cotton (96), canola (93), sugarbeet (95), alfalfa (30), papaya, squash
Brazil	99.6	Soybean, corn, cotton
Argentina	60.3	Soybean, corn, cotton
India	27.2	Cotton
Canada	26.7	Canola (96), corn, soybean, sugarbeet
China	10.4	Cotton, papaya, poplar, tomato, sweet pepper
Paraguay	8.9	Corn, cotton, soybean
South Africa	7.2	Corn, cotton, soybean
Pakistan	6.9	Cotton
Uruguay	3.7	Corn, soybean
19 other countries grow less than 2 million acres.		

[a]Number in parentheses is % of total crop area.

of a population.[10] The affected population and the control are nearly impossible to create, which is necessary for a proper experiment. In addition, we do not know how long a long-term study must be to determine possibility of the object of the study—the, expected, known, and unexpected unknown effects.

There have been several economic and practical benefits from the use of GE in agriculture. Those most frequently mentioned, but not agreed on, include increased crop yield, reduced costs, reduced need for pesticides, increased resistance to pests and diseases, and greater food security. A strong voice that disputes the claim of no human harm is Vandana Shiva, an Indian environmental activist and antiglobalization author. She, with a PhD in the philosophy of science from the University of Western Ontario, brings scientific credibility to the debate. She has received numerous international awards, and is a leader in the Indian environmental movement. Her opposition to GM crops has made her a hero to anti-GMO activists (Specter, 2014). She claims that most of today's GM crops have been developed by large corporations in the industrial West to meet the needs of farmers in developed countries. She argues that while the benefits

---

10. Personal communication, Thompson, P., Professor of Philosophy, Michigan State University. December 2014.

TABLE 8.2 Some Genetically Engineered[a] Plants

Food/Feed/Fiber Crops (James, 2013)

Alfalfa

Argentine

Canola

Bean

Cotton

Chicory

Corn (animal and sweet)

Eggplant

Flax

Melon

Papaya

Plum

Polish canola

Poplar

Potato

Rice

Soybean

Squash

Sugarbeet

Sugarcane

Sweet pepper

Tobacco

Tomato

Wheat

Other Plants

Carnation

Chicory

Creeping

Bentgrass

Petunia

Rose

[a]http://en.wikipedia.org/wiki/Genetically_modified_crops.

TABLE 8.3 Some Countries Where GM Crops Have Been Planted and Total Approvals of 100+ from 1992 to 2014 (James, 2013)

Argentina (112)
Australia (122)
Bangladesh
Bolivia
Brazil (115)
Burkina Faso
Canada (344)
Chile
China (100)
Colombia (110)
Costa Rica
Egypt
European Union (5 countries: 142)
Honduras
India
Indonesia
Iran
Japan (475)
Malaysia
Mexico (159)
Myanmar
New Zealand
Norway
Pakistan
Panama
Paraguay
Philippines (161)
Russia
Singapore

TABLE 8.3 Some Countries Where GM Crops Have Been Planted and Total Approvals of 100+ from 1992 to 2014 (James, 2013) — cont'd

South Africa (117)
South Korea (200)
Switzerland
Taiwan
Thailand
Turkey
United States (440)
Uruguay
Vietnam

http://en.wikipedia.org/wiki/Genetically_modified_crops.

mentioned previously may be true for large, industrial farmers, they are not true for the thousands of small farmers in India. She is a formidable opponent, but many dismiss her as an idealist who wants to return to the agriculture of the past, which they claim cannot feed the world.

Agricultural Goals of Recombinant DNA Technology

Drought and heat (stress) resistant crops
Crop resistance to insects, plant pathogens, and viruses
  Resistance to plum pox virus has been achieved by insertion of a coat protein gene from the virus (Scorza et al., 2013).
Crop resistance to one or more herbicides (achieved)
Management of parasitic weeds
  Tropical corn is plagued by root parasitic witchweeds (*Striga* spp.), which are not controlled by selective herbicides (Gressel, 1999). Corn seeds dressed with one of two herbicides (imazapyr and pyrithiobac) provide season-long control of witchweed in imidazolinone-resistant corn.
Crop plants that produce their own insecticide (achieved)
  Bacillus thuringiensis (Bt) is a naturally occurring soil bacterium that produces a protein that kills the European corn borer and cotton bollworm and budworm. Corn and cotton have been genetically engineered to contain the Bt toxin.
Improved crop yield
Improved crop response to fertilizer that may reduce the need for fertilizer
Enhanced flavor and nutritional content of vegetables and fruits

Soybean (Kinney, 1994).

Canola (Abidi et al., 1999). The saturated fatty acid laurate content of canola was increased by insertion of a gene from the California bay tree (*Umbellularia californica*) to make the oil more suitable for industrial uses (cosmetics, soap, detergents) where imported palm and coconut oil had been used.[11]

Delayed fruit ripening for longer shelf life during transportation

Edible vaccines to prevent widespread diseases in developing countries

Production of biofuels

Production of useful by-products (biopharming)—for example, pharmaceutical drugs, natural materials(bioplastics)

Phytoremediation—the use of transgenic plants and their associated microbes to detect contaminants in soils, sediments, and groundwater that sense and metabolize explosives and other contaminants; for example, *Arabidopsis thaliana* (mouse ear-cress) has been used to detect the explosive Royal Demolition Explosive-cyclotrimethylenetrinitramine $(C_3H_6N_6O_6)$[12] and some heavy metals. Tobacco has been modified to detect TNT (trinitrotoluene, $C_6H_2(NO_3)CH_3)$ (Haselhorst, 1999; Singh and Mishra, 2014). An excellent review of the topic is available (Cherian and Oliveira, 2005)

Human Health Goals of Recombinant DNA Technology

Recombinant Vaccines (i.e., hepatitis B)[13]

Prevention and cure of sickle cell anemia

Prevention and cure of cystic fibrosis

Production of clotting factors

Production of insulin

Production of recombinant pharmaceuticals—This is clearly a benefit to human health, but because the medicinal products will be produced in GM agricultural crops it will inevitably be controversial.

Germ line and somatic gene therapy

THE CASE AGAINST USE OF RECOMBINANT DNA TECHNOLOGY IN AGRICULTURE

In her books (1993, 1997, 2000) and in her public presentations, Shiva presents the case against genetic modification. Jensen (2013) paints an equally dark picture. Capitalism is a predictable process whereby dangerous technologies (pesticides, bovine growth hormone, nanotechnology, nuclear power, and, of course, GMOs) are introduced, become widespread, and therefore, it is claimed, are

---

11. http://www.smallgrains.org/Springwh/March97/Canola.htm. Accessed June 2014.

12. See http://www.isaaa.org/resources/publications/pocketk/25/default.asp. Accessed June 2014.

13. http://en.wikipedia.org/wiki/Hepatitis_B_vaccine Accessed June 2014.

uncontrollable and will inevitably be harmful. He claims that when a brilliant tactical solution to an important problem is proposed it is inevitable that those many accuse of being luddites, malcontents, or pessimists will object and begin to do battle with those who propose, develop and, one assumes, will profit from the new technology.

It is reasonable to claim that the case against widespread use of recombinant DNA technology and the other agricultural technological advances Jensen (2013) mentions is frequently based on suppositions, irrational fear (Van Montagu, 2013), and about what might, but has not happened. That is they are not objections based on scientific evidence and, in fact, frequently ignore or reject scientific evidence (see Bronner, 2014; Seidler, 2014). Objections are commonly based on popular beliefs. But, beliefs are just that: beliefs. Although that may be true it does not permit ignoring or automatically dismissing what the public believes. Beliefs are powerful creators of public opinion that, if ignored, imperils public acceptance, potential benefits, the developer's reputation and, perhaps, eventual profit.

There are several examples of the force of public opinion.

1. A lawsuit has been initiated on Kauai, Hawaii, to make the case against a legislative ban on growing GM crops on the island (D'Angelo, 2014).

Maui County voters opted to temporarily ban cultivation of genetically engineered crops (50% yes, 48% no). In 2013, Hawaii County prohibited GE farming, with the exception of papayas. Kauai County passed a groundbreaking GMO bill that limits biotech crop and pesticide use. Three of the world's largest agrochemical companies (Agrigenetics, Syngenta, and DuPont) sued in Kauai, Maui, and Hawaii to overturn the legislative initiatives. The laws were declared unconstitutional in late November 2014.

2. Vermont's legislature's passed a labeling law for GM foods, which will be effective in 2016 (Economist, 2014a). Connecticut and Maine have passed labeling laws, but both states stipulate that the law cannot become effective until other states have passed similar laws.
3. Colorado's citizen-initiated proposition 105, on the 2014 ballot, sought to mandate labeling of GMOs except meat, dairy products, alcoholic beverages, and restaurant food. There was vigorous opposition (CAST, 2014; Entine, J.[14]); it failed 2:1. A similar ballot issue in Oregon also failed 50.5 to 50.
4. China rejected GMO corn in distiller's grains imported from the United States (Huff Post Green, 2014b).
5. There were fierce public protests against GM food in China (Economist, 2013).
6. Vigorous objection was made to the US Department of Agriculture's acceptance of GM soybean capable of resisting the herbicide 2,4-D. The objection

---

14. http://www.forbes.com/sites/joentine/2014/08/25/why-liberal-americans-are-turning-against-gmo-labelling/.

was based on the fact that 2,4-D was a component of Agent Orange during the Vietnam War.

A widely debated application of recombinant DNA technology is production of GM foods that, it is claimed, may negatively affect human health. Genes can be derived from unrelated plants or other organisms to give plants characteristics that proponents argue are beneficial to both producers and consumers of agricultural products. For most engaged in scientific research and those who create and market GM foods, the potential benefits of recombinant DNA significantly outweigh concern, especially what is regarded as irrational concern, about public health. The potential for future unknown, detrimental risks (allergies, negative environmental effects) from the source of DNA, are nonissues because scientists and those who market food are neither irrational nor stupid. There is concern about future negative outcomes in the scientific community and among the developers and users of GE technology.

The history of scientific successes and disasters is readily available; some emphasize the former, some the latter. The debate within and outside the agricultural community over the risks and ultimate beneficiaries of genetic modification of crops has raised legitimate biological, economic, moral, and social concerns. If one views science and technology over the past 100 years, it is clear the optimists have had it right about the benefits of agricultural technology (Tenner, 1996). Nevertheless, public fear is real, not irrational. People are regularly notified of fallible technology and agriculture has not escaped public scrutiny (see Zimdahl, 2012; Chapter 3).

Agricultural Examples of Fallible Technology

- The mid-1960s controversy over the real and suspected hazards of the dioxin contaminant of 2,4,5-T, a component of Agent Orange, used in Operation Ranch Hand, a vegetation control program during the Vietnam war.
- On December 3, 1984, a poisonous cloud of methyl isocyanate, used in the manufacture of pesticides, escaped from Union Carbide's plant in Bhopal, India, killing 14,000 and permanently injuring 30,000 people.
- No one knows for sure, but in 1984 it is estimated that between 1 and 5 million cases of pesticide poisoning occur every year in the world, resulting in 20,000 deaths. In 1990, the estimate was 1 million unintentional deaths, 2 million deaths by suicide, and 25 million episodes of poisoning (Jeyaratnam, 1990). Developing countries use 25% of pesticides and experience 99% of the deaths (Goldman, 2004).
- Air and water pollution and animal suffering from confined animal feeding operations.
- Mad cow disease, swine flu, bird flu, meat recalls, and antibiotic resistance (see Chapter 9) are all of concern or have been major societal concerns in the past.
- The contributions of nitrogen fertilizer to the ecological dead zone extending into the Gulf of Mexico from the Mississippi river's terminus.

Other Examples of Fallible Technology

- On April 26, 1986, unit four at the nuclear power plant in Chernobyl, Ukraine, exploded, causing approximately 6000 deaths and injuring more than 30,000 people.
- On January 28, 1986, 73 s into its tenth flight, the space ship *Challenger* exploded, killing seven astronauts.
- The 1989 spill of 11 million gallons of oil from the Exxon Valdez.
- In 2005, Dell recalled 4.1 million notebook computers when it was discovered that their batteries were a fire risk.
- The failure of more than 50 levees in August 2005 in and around New Orleans, LA, during Hurricane Katrina.
- The collapse of the eight-lane, steel-truss bridge on Interstate 35 over the Mississippi River on the evening of August 1, 2007.
- On August 10, 2010, the Deepwater Horizon drilling platform caught fire in the Gulf of Mexico. There have been several other oil spills.
- The continuing fear and uncertainty of the nuclear crisis at the Fukishima Daiichi nuclear plant in Japan that began after the earthquake and tsunami in March 2011.

The general public is aware of these and other examples of scientific fallibility. It is quite possible they are also aware of and their concern is increased by scientific uncertainty. A few examples of conflicting scientific studies illustrate the point (Reynolds, 2014).

- One study in humans reported that if one sleeps less than 6 h/night, they may die younger. Another reported that people who sleep more than 8 h may die younger.
- Blueberries extended the life of fruit flies, but another study reported that blueberries had no effect on life span of mice.
- Rapamycin (an antibiotic) did not slow aging of mice in one study but it did in another.
- Being slightly overweight did not decrease life span in humans in one study, but in another being underweight did.

Reynolds notes that her article is an overview of the ups, downs, reverses, and highlights of recent research. Scientists will not be surprised. It is the very nature of science that disagreements are inevitable and will be resolved with further experiments.

The agricultural view has been that the examples are true and what was reported, happened. But the view always includes the question of whether or not the particular agricultural technology or practice is really dangerous and doomed to fail. Of course, the answer depends on how danger is defined. Nearly everything can be toxic if enough is consumed or exposure is very high. Carbon monoxide is present on city streets and, although automobile exhaust is unpleasant, exposure while stuck in traffic is not harmful.

Its toxicity is easily experienced when one sits in a running car in a closed garage. The dose makes the difference. An ancillary aspect of public concern is that independent of the event there is rarely an ameliorative technology on a par with what has failed. The counsel of the English writer and dramatist Douglas Adams (2002, p. 720) is worthy of consideration:

> *The major difference between a thing that might go wrong and a thing that cannot possibly go wrong is that when a thing that cannot possibly go wrong goes wrong it usually turns out to be impossible to get at or repair.*

On the other hand, there is a plethora of examples that affirm that agricultural science and technology have improved food production and human health and made life better. Disasters have occurred but they "mobilize the kind of human ingenuity that technological optimists believe exists in unlimited supply" (Tenner, 1996). Technological disasters will continue to occur and agriculture will cause some, but, on balance, because of good science and concerned agricultural scientists, the quality and quantity of the world's food will continue to improve for most people.

Similarly many, Shiva among them, argue that possible environmental issues associated with transgenic crops have not been fully resolved and should be before widespread use. This is also an important issue that must be continually addressed, but the concern ignores overwhelming scientific evidence about the lack of environmental or human harm. In short, concern is not based on available scientific evidence. But that is central to the concern. It is not that there is no scientific evidence of harm. It is, as pointed out previously (Thompson, 2007), that no one has looked for evidence of harm. Proponents claim that agricultural biotechnology offers an unparalleled safety record as well as widespread commercial acceptance by farmers. Those who make a case against agricultural biotechnology argue that those in favor are focused primarily on profit. They ignore two things. First, profit is necessary, not, per se, evil. The second is that they ignore the overwhelming evidence of safety.

A major concern for those opposed to biotechnology is based on human health and choice. They acknowledge that 60–70% of processed food in the grocery store contains GM ingredients that are not labeled. One of the tenets of a democratic society is consent of the governed. Because food is essential to life, opponents claim, people should be given a choice of whether they want to consume food that contains GM ingredients. CAST (2014) argues that people do have a right to know what is in their food, but it does not logically follow that they have a right to know how their food is produced. Their claim is that the use of the right to know argument "uniquely singles out GE technology." There are more than 400 items (baby food, cereals, meat, and sodas) in the grocery store that are GM or contain a GMO. It is quite reasonable to claim that during the nearly 20 years that GM food has been available, people, at least those in developing countries, have consumed billions of meals that contained some GM ingredients. The CAST report (2014) argues that not one documented health or disease problem has occurred. Nevertheless, the ideological and perhaps political question of labeling continues (e.g., the ballot issues mentioned previously). Rather than being based on scientific evidence, they are derived from a host of fears and possibilities (e.g., possible human health hazards, environmental risks, the right to know, crossing forbidden species boundaries, corporate profiteering, Western colonialism) (Van Montagu, 2013).

Recombinant DNA technology has contributed important benefits for agriculture and human health. More are promised. Noble (2013) points out that the rapid scientific progress has created a new risk. The research and knowledge gained could be misappropriated for harmful purposes. "Synthetic biology can help produce better medicines and cleaner manufacturing processes, but in criminal hands, it could also be used intentionally to modify an existing disease or create a novel, highly pathogenic biological agent." There is no coordinated international response or regulation of synthetic biology by governments or the scientific community. He says that because the potential for misuse is so great, regulation is required. However, Noble also argues that science should proceed unhindered. The potential benefits of synthetic biology research are so great that overregulation may be as great a threat as no regulation.

PRESENT AGRICULTURAL DEVELOPMENTS

Viral Ringspot Resistant Papaya

Dr Dennis Gonsalves (1998, 2004) successfully developed transgenic papaya resistant to the pathogenic plant virus ringspot, which was first found in Hilo, Hawaii, in 1970. Gonsalves, a Hawaiian native and plant pathologist at Cornell University, is credited with bringing the Hawaiian papaya industry back from the brink of extinction. In the 1990s, most of the papaya trees and the industry on Hawaii (–"The Big Island") were in extremis; nearly all the trees were dead or dying.

Because the resistant genes exist only in the seed, not in the fruit, regulatory agencies (US Environmental Protection Agency, FDA) found the transgenic fruit to be substantially equivalent (a term not accepted by opponents as a synonym for "the same") to nontransgenic (GM) papaya. No evidence of significant dietary or allergenic differences was found in a 2011 study by scientists from the Pacific Basin Agricultural Research Center in Hilo and the University of Hawaii. Evidence that Thompson's (2007) claim that no one has looked may be wrong. Gonsalves believes health concerns are unwarranted.[15]

In spite of the significant benefits to the papaya industry, objections to genetic modification of crops in Hawaii are gaining ground (see previous discussion) (Harmon, 2014). A wide array of issues is cited. Many people object to

---

15. http://hawaiitribune-herald.com/sections/news/local-news/papaya-gmo-success-story.html#sthash. dJ0RI4ob.dpuf. Accessed May 2014.

consuming genetically altered foods. Others question the validity of studies on the risk of consuming GM, especially over the long term. Some object to what they see as corporate greed by biotech companies and their quest for control of the world's food supply (Huff Post Green, 2014a).

Golden Rice

Another significant achievement is the development of Golden rice, a recombinant variety of rice that was engineered to express the enzymes responsible for biosynthesis of β-carotene, the precursor of vitamin A, in the edible parts of rice. It is a product of insertion of two carotenoid biosynthesis transgenes phytoene synthase from daffodil (Narcissus) and the bacterial carotene desaturase (lycopene), which is red. Thus, the endosperm of Golden rice is yellow because of the accumulation of β-carotene (provitamin A) and xanthophylls.

The United Nations Special Session on Children in 2002 set the elimination of vitamin A deficiency as a goal for 2010. It has not happened. The Golden rice research goal, consistent with the United Nations goal, was to produce a rice variety that could be grown and consumed in areas with a shortage of dietary vitamin A. Its deficiency is estimated to affect 190 million preschool children and 19 million pregnant women annually and kill 670,000 children younger than age 5 each year (WHO, 2009). Approximately 250,000–500,000 malnourished children younger than age 5 go blind each year from vitamin A deficiency, and approximately half die within a year of becoming blind. The deficiency is common in poorer countries and rarely seen in more developed countries. Night blindness is one of the first signs of vitamin A deficiency. Approaches to addressing the problem (e.g., vitamin A supplementation, food diversification, food fortification) have been successful, but millions still suffer, suggesting that food/aid distribution is missing populations in remote areas. Because rice is widely produced and consumed, Golden rice has the potential to help people. Its distribution and availability will complement current efforts.

Golden rice was developed in an 8-year project by Ingo Potrykus of the Swiss Federal Institute of Technology in Zurich and Peter Beyer of the University of Freiburg in Germany. The achievement was applauded when Potrykus appeared on TIME's July 31, 2000, cover. The scientific details were published in *Science* in 2000 (Ye et al.). Within the scientific community, it is regarded as a significant breakthrough in biotechnology, because an entire biosynthetic pathway had been engineered. Golden rice-1, the product of the research by Potrykus and Beyer, contains approximately 1.6 μg of carotenoids, which equals 0.8 μg of β-carotene/g of dry rice. It was criticized because one would have to eat a very large amount to receive significant benefit, which made it highly likely that no benefit would result.

In 2005, scientists developed the current version of Golden rice, GR2-R, using genes from corn and a common soil microorganism that together produce up to 23 times more β-carotene than the original prototype (Paine et al., 2005).

It holds substantial promise for reducing the incidence of vitamin A deficiency in the world. Daily consumption of about a cup (or around 150 g uncooked weight) could supply 50% of the Recommended Daily Allowance of vitamin A for an adult (Tang et al., 2009).

A group of Australian scientists at the Queensland University of Technology in Brisbane have taken a different route to address vitamin A deficiency by genetically modifying bananas to increase their vitamin A content as much as five times (TIME, 2014b).[16] The Australian bananas will undergo human testing in India. The approach to adoption is unique. Village leaders in Africa will be given 10 free vitamin A–enriched banana plants on the condition that they give at least 20 new shoots to other villagers (TIME, 2014a). It is a pragmatic, demonstrative approach rather than the more traditional oral approach.

Although Golden rice and vitamin A–enriched bananas were developed as humanitarian tools, both have encountered significant opposition. Shiva is opposed to Golden rice (see Specter, 2014, p. 56). Currently, development and evaluation are continuing. Widespread use depends on resolution of regulatory issues. Golden rice will be available to farmers and consumers when it has been successfully introgressed[17] into rice varieties suitable for Asia after its safety and efficacy to improve vitamin A status in communities have been proven, and it has been approved by national regulators.

It is worth noting that the initial research by Potrykus and Beyer and subsequent more extensive research by scientists at the International Rice Research Institute on Golden rice has all been in the public realm, rather than by for-profit, multinational organizations. This is also true for the Australian research on bananas. International Rice Research Institutes research is a significant part of what many call the second green revolution, which focuses on rice. Roughly a third of humanity depends on rice for 50% of their daily caloric intake. It is the most important world food crop in terms of the number of people who depend on it every day.

The first green revolution was largely government-backed with significant help from the international agricultural research centers and American foundations (i.e., Ford and Rockefeller). The second green revolution will not depend upon a few miracle varieties as the first did (Economist, 2014b). It will depend upon scientist's ability, using traditional plant breeding and GE, to tailor existing crop cultivars, to different environments. That work is being done, primarily, in public research institutions (i.e., universities and international agricultural research institutes).

---

16. http://www.independent.co.uk/news/science/gm-banana-designed-to-slash-african-infant-mortality-enters-human-trials-9541380.html. Accessed June 2014.

17. Introgression: a long-term process in which a gene from one species is moved, using traditional plant breeding techniques, into the gene pool of another by repeated backcrossing of an interspecific hybrid with one of its parents.

TABLE 8.4 Number of GM Events[a] for Some Crops

Crop	Number of Events
Canola	34
Corn/maize	135
Cotton	52
Potato	31
Rice	7
Soybean	30
Sugarbeet	3
Tobacco	2
Tomato	11
Wheat	1
Total	306

[a]An event is a specific genetic modification in a species. Table 8.4 shows only some of the 3059 genetically modified events approved in 38 countries by October 2014. http://www.isaaa.org/gmapprovaldatabase/eventslist/default.asp. Accessed October 2014.

Herbicide-Resistant Crops

Several crops have been genetically modified. Table 8.4 shows the number of GM events of all kinds approved for several crop. Monsanto's Round-Up Ready™ corn was first approved for use in 1996. Since then, several crops have been genetically engineered to be resistant to one or more herbicides (Table 8.5). When recombinant genes for different traits are combined in a single crop, the process is called stacking. Incorporation of herbicide resistance alone simplifies weed control because a herbicide, which would normally kill the crop, can be used to achieve effective weed control. Stacking herbicide resistance with the natural biological insecticide Bt.

B. thuringiensis provides insect control in some crops and weed control when the proper herbicide is applied. Herbicide resistance has been achieved in 28 crops. Stacked traits were included in crops planted in 13 countries, which planted 27% of the world total of 432 million acres of biotech crops in 2013 (James, 2013).

The economic benefits of GM crops and the extent of their adoption by farmers virtually assures continued development of weed resistance to more herbicides in more crops. GE of agricultural crops will increase. There are several potential benefits, although few have been achieved: increased yield, reduced production costs, reduced need for pesticides, and increased food

TABLE 8.5 Some Crops That Have Been Genetically Modified to Be Resistant to One or More Herbicides

Herbicide	Resistant Crops
Alfalfa	Glyphosate (Round-Up™ and others) and glufosinate (Finale™ and others)
Canola/rapeseed	Glyphosate and glufosinate
Corn	Glyphosate, glufosinate, some sulfonylureas and imidazolines, and 2,4-D
Cotton	Bromoxynil, sulfonylureas
Flax	Sulfonylureas
Soybeans	Dicamba, glyphosate, glufosinate, some sulfonylureas and imidazolinones, and 2,4-D
Sugar beet	Glyphosate and glufosinate
Wheat	Glyphosate

production. It is generally agreed that crop yield has not gone up because of improvements in a crop's productive potential. Yield increases have been due to more careful crop management and improved weed and insect control.

A frequent criticism of the widespread development and use of herbicide-resistant crops is the inevitable emergence of what are called superweeds. One must admit that several weeds have characteristics (e.g., abundant seed production, adaptation to growing in cultivated crops, perennial growth, ability to distribute seeds widely) that make them seem especially pernicious. They are very good weeds, but they are not superweeds. The term "super" has become a synonym for "cannot be controlled." It is clear, and seemingly always has been, that it is extremely difficult to selectively control perennial weeds (field bindweed, leafy spurge, and several others) with herbicides or any one of several other techniques. It is possible and relatively easy to control some perennials (e.g., dandelions), most annual weeds, and some biennials with one application of an appropriate herbicide or with a hoe. They are not super. What has happened is continued use of an herbicide over several years on a GM crop to be immune to the effects of the herbicide has resulted in widespread, inevitable development of resistance to the herbicide by the weed(s) it had controlled: which means that an increasing number of weedy species that had been, are no longer controlled. However, these weeds may be less fit, in the ecological sense than their non-GM relatives and thus less able to adapt to crop environments, less competitive, and/or less invasive. That is, their weediness may be lessened.

Similarly, some claim that crops that have been engineered to be resistant to or tolerant of one or more herbicides might crossbreed naturally with related plant species and transfer the herbicide-resistant genes to the related species, which could then become superweeds. Gene transfer is a definite possibility. But recipient plants will not be endowed with supernatural powers of survival. Because of the absence of close genetic relatives the possible transfer of a resistant gene to a weed or close, nonweedy relative, is remote, if not impossible. Transfer of a genetic modification, known as gene flow (e.g., herbicide resistance) from oats (*Avena sativa* L.) to wild oats (*Avena fatua* L.) or from canola (*Brassica napus* L.) to wild mustard [(*Brassica kaber* (DC)L.C., Wheeler)] or other wild *Brassica* species would make chemical control of the now herbicide-resistant weed much more difficult, if not impossible. The same possibility is apparent for sugarbeet with wild beet, squash, other Cucurbits (e.g., cucumber, gourds), and sunflower with wild sunflower.

The phenomenon and development of herbicide resistance has been observed for at least 50 years. However, during the past few decades, coincident with the advent of herbicide-resistant crops, repetitive use of the same herbicide on the same crop in the same field has become common and the development of weed resistance has increased dramatically. Fewer than 10 resistant weeds were known in 1955, whereas in October 2014 there were 435 documented cases of weed resistance. An exponential increase in weed resistance to glyphosate (Round-Up) began in 1975 and in 1990 for acetolactate synthase inhibitors (sulfonylureas and imidazolinones). Acetolactate synthase is the enzyme responsible for synthesis of the amino acids valine, leucine, and isoleucine (Heap, 2014). Resistance is a significant weed management problem and the focus of a great deal of scientific attention. In November 2014, there were 437 unique cases of herbicide resistance in the world among 238 different weed species (138 dicotyledons and 100 monocotyledons). There are weeds that are resistant to 22 of 25 known herbicide modes of action. There are documented cases of resistance to 155 different herbicides in 84 crops and 65 countries (Heap, 2014; November 12).

There have been changes in herbicide use. Overall, because of the increased use of glyphosate and glufosinate resistant crops (see Table 8.5), more of these herbicides have been used. One can make the reasonable claim that the use of other, perhaps less environmentally desirable, herbicides has diminished. But, total herbicide use has not decreased because of crop GM. Benbrook (2009, 2012) supports that claim. The 2009 report presents data to show that herbicide-tolerant crops increased herbicide use by 383 million pounds during 13 years of commercial use, whereas insecticide use declined 64.2 million pounds. The 2012 (1966–2011) report claims an increase of 527 million pounds of herbicides, and a continuing decrease in insecticide use (primarily from Bt cotton and corn) of 64.2 million pounds. It is a bit ironic that in increased herbicide use has exacerbated (caused) weed resistance and made weed control a serious, challenging, and increasing problem in GM crops (especially corn and soybeans).

Insect-Resistant Crops

B. thuringiensis is a bacteria that naturally produces a protein (Bt toxin) with insecticidal properties. More than 100 different chemical variations of Bt toxin have been identified in different strains of the bacteria. The toxin has been widely adopted in agriculture and gardening as an insect control strategy for at least 50 years. Beginning in 1996, Bt corn that expressed a recombinant form of the bacterial protein, which effectively controls some insect predators, became available. Bt corn has been genetically altered to express one or more proteins from the *B. thuringiensis* bacteria. The Bt protein is expressed throughout the GM plant. When a vulnerable insect eats the Bt-containing plant, the protein is activated in its alkaline gut (the human gut is acidic). In the alkaline environment, the protein partially unfolds, is enzymatically degraded, forms a toxin that paralyzes the insect's digestive system, and creates holes in the gut wall. The insect stops eating within a few hours and eventually starves. Bt corn has reduced the billions of dollars lost annually from the European corn borer, corn ear worms, corn root worms, and the cotton bollworm each year. Plants genetically modified to express other proteins toxic to insects are being developed.

Innate™ Potato

In late 2014, The US Department of Agriculture deregulated potato varieties created through a proprietary biotechnology process called Innate. The technique developed by the J.R. Simplot Co. achieves desirable traits without incorporating any foreign genes. The Innate potatoes will have about 40% less bruising from harvest impacts and storage pressure than conventional potatoes and have lower levels of asparagine, which frying converts to acrylamide, a chemical linked to cancer. Simplot predicts Innate potatoes will reduce waste by 400 million pounds in the food service and retail industries and a significant portion of the billion pounds discarded by consumers. Further the biotech potatoes pose no environmental risk, create no harm to other species, and grow just like conventional potatoes.

FUTURE EFFECTS

The five examples of successful application of biotech crop/DNA technology in the preceding section are now part of the agricultural enterprise. They have been accomplished. Each has or soon will achieve widespread acceptance within the agricultural community and each has resulted in serious, thoughtful objections, and challenges by others. Given the public and proprietary scientific research on genetics and biotechnology, it is reasonable to claim that most of the present agricultural and human health goals of recombinant DNA technology (see Agricultural goals of recombinant DNA technology and Human health goals of recombinant DNA technology) will be achieved in the twenty-first century. For example, by 2030, it is likely that rice scientists will have produced a more robust C_4 and nitrogen-fixing rice varieties and more nutritious rice (Anonymous, 2014). Much of agriculture's scientific foundation will remain, but the application of recombinant DNA will change many aspects of agricultural practice and its products. Scientists will achieve recombinant DNA's present goals and move on to new, perhaps presently unknown research goals. Agricultural scientists, commercial interests, and producers will readily adopt the new technologies and, when challenged, will justify adoption by invoking their economic efficiency, potential to increase yield and agriculture's generally agreed upon moral justification: they will help feed the world.

The rate of world population growth is slowing. It peaked at 2.2% in 1963, declined to less than 1.1% in 2012, and is forecast to be as low as 0.5% by 2050. Nevertheless, agriculture's challenge remains because world population will continue to increase beyond the present 7.2 billion. The United Nations' pessimistic forecast is 26 billion by 2050. The median forecast is 10 and the lowest is a decline to 6 billion by 2050. Unless people and their governments agree on how to limit population, agriculture's practitioners must struggle to produce more. Biotechnology is viewed by many as the best of the available ways to increase productivity, protect the environment, and ensure safe food for all. Biotechnology's agricultural goals (see Agricultural goals of recombinant DNA technology) are designed to accomplish these things.

It is doubtful that those opposed to biotechnology are also opposed to agriculture's goal of feeding the world. It is generally agreed that it is a good, ethically correct, thing to do. Thus, one must ask, as Thompson (2007) does, why is there such strong opposition. He suggests three reasons.

1. Manipulation of genes, the foundation of much of biotechnology, is inherently wrong. It is absolutely forbidden.
2. Some (not all) of the present achievements of biotechnology are based on scientific research done in the public realm, but it appears the benefits are captured in the private realm of corporate power within the capitalistic society.
3. A third, perhaps less persuasive, argument is often disguised under the heading of botanical prospecting. Scientists from developed countries discover indigenous people who have developed medical or other benefits from plants or their products. The botanical prospector takes and uses the genes (germplasm) of the plants to develop new medicines or use genes to create GM crops, which can be patented with the result that the knowledge (the intellectual property rights) of the indigenous people are, or can be, easily ignored.

Thompson (2007, p. 63) states that "biotechnology's boosters" have "done serious damage to their own case by offering several singularly bad arguments." Thompson deals with them at length, which is beyond the scope of this book. The four fallacious arguments he describes are as follows.

1. The modernist fallacy appeals to an "outdated and naive notion of technological progress." The notion is that ethical concerns are simply the price of progress and dwelling on them may deter progress. Scientific and technology progress has always created changes, most of which have been beneficial; let's emphasize the positive.

2. The naturalistic fallacy uses an inappropriate reference group to make comparisons of the relative risks of biotechnology. It is the claim that if something is natural it must be good. Thompson offers a relevant example: "The kinds of alterations that molecular biologists are making in plants and animals are just like those that occur as a result of natural mutation. They are, therefore, an acceptable risk."

3. The argument from ignorance is an especially troubling accusation. Ignorance proclaims that because there is no evidence of harm from recombinant DNA products that have been on the market for many years, no harm is likely to occur.

4. There is inappropriate emphasis on biotechnology as the next great advance in agriculture, the foundation of the next green revolution, and, thereby, the solution to world hunger. Those whom Norman Borlaug, the father of the first green revolution and Nobel Prize Peace laureate, called "anti-science zealots" simply don't understand agriculture's goal and the essentiality of biotechnology to achieving it. In fact the zealots are "ethically irresponsible" because their opposition to progress will perpetuate and increase hunger, starvation, disease, and poverty.

If agriculture's practitioners continue to emphasize only the positive benefits of their scientific achievements and the promises of biotechnology, while dismissing or ignoring the moral arguments, they will delay their progress and risk not fulfilling their mission. Agriculture's reliance and continued quest for increasing production as the justification for all activities is understandable and risky. Agriculture's paradigm is shifting.

The chemistry of biotechnology is sophisticated and often seems omnipresent in the agricultural world. In spite of serious moral and scientific challenges, it is the wave of the future of production agriculture. Its myriad benefits must be made clear to opponents who concurrently are obligated to listen and responsibly debate the positive and negative aspects of the technology.

REFERENCES

Abidi, S.L., List, G.R., Rennick, K.A., 1999. Effect of genetic modification on the distribution of minor constituents in canola oil. Journal of the American Oil Chemists' Society 76 (4), 463–467.

Adams, D.N., 2002. The Ultimate Hitch Hikers Guide to the Galaxy—Mostly Harmless. Random House Publishing, New York. 815 pp.

Anonymous, 2014. Researchers Rally for Rice Science at IRC2014. Crop Biotech Update. International Service for the Acquisition of Agri-Biotech Applications (ISAAA). See: cropbiotechupdate=isaaa.org@maqil68.wdc03.rsgsv.net.

Avery, O.T., Macleod, C.M., McCarty, M., 2000. Studies on the chemical nature of the substance inducing transformation of pneumococcal types: induction of transformation by a desoxyribonucleic acid fraction isolated from pneumococcus type III. Clinical Orthopaedics and Related Research 379 (Suppl.), S3–S8.

Benbrook, C., 2009. Impacts of Genetically Engineered Crops on Pesticide Use in the United States: The First Thirteen Years. Critical Issue Report. The Organic Center, Washington, D.C. 62 pp.

Benbrook, C., 2012. Impacts of genetically engineered crops on pesticide use in the United States: the first sixteen years. Environmental Sciences Europe. 24:24. http://www.enveurope.com/content/24/1/24.

Bonner, D., 2014. Herbicide and Insecticide Use on GMO Crops Skyrocketing While pro-GMO Media Run Interference.

CAST–Council for Agricultural Science and Technology, 2014. The Potential Impacts of Mandatory Labelling for Genetically Engineered Food in the United States. Issue Paper 54. CAST, Ames, IA.

Chargaff, E., Zamenhof, S., Green, C., 1950. Composition of human desoxypentose nucleic acid. Nature 165 (4202), 756–757.

Chargaff, E., Lipshitz, R., Green, C., 1952. Composition of the deoxypentose nucleic acids of four genera of sea-urchin. Journal of Biological Chemistry 195 (1), 155–160.

Cherian, S., Oliveira, M., 2005. Transgenic plants in phytoremediation: recent advances and new possibilities. Environmental Science and Technology 39, 9377–9390.

D'Angelo, C., February 28, 2014. Honolulu firm to Defend County. The Garden Island. P. A-1, A-8.

Dahm, R., 2008. Discovering DNA: Friedrich Miescher and the early years of nucleic acid research. Human Genetics 122 (6), 565–581.

Economist, December 14, 2013. Food Fight. pp. 53–54.

Economist, May 10, 2014a. Vermont v Science. pp. 25–26.

Economist, May 10, 2014b. A Second Green Revolution. p. 14.

Elson, D., Chargaff, E., 1952. On the deoxyribonucleic acid content of sea urchin gametes. Experientia 8 (4), 143–145.

Garrett, L., November/December 2013. Biology's Brave New World—The Promise and Perils of the Synbio Revolution. Foreign Affairs. pp. 28–53.

Goldman, L., 2004. Childhood Pesticide Poisoning: Information for Advocacy and Action. United Nations Environment Programme, Châtelaine, Switzerland. 37 pp.

Gonsalves, D., 1998. Control of papaya ringspot virus in papaya: a case study. Annual Review of Phytopathology 36, 415–437.

Gonsalves, D., Gonsalves, C., Ferreira, S., Pitz, K., Fitch, M., Manshardt, R., Slightom, J., July 2004. Transgenic Virus Resistant Papaya: From Hope to Reality for Controlling Papaya Ringspot Virus in Hawaii. American Phytopathology Society Net.

Gressel, J., 1999. Herbicide resistant tropical maize and rice: needs and biosafety considerations. The 1999 Brighton Conference—Weeds, vol. 2, British Crop Protection Council, pp. P637–P645.

Harmon, A., January 11, 2014. A Lonely Quest for Facts on Genetically Modified Crops. NY Times. http://www.nytimes.com/2014/01/05/us.on-hawaii-a-lonely-quest-for-facts-about-gmos.html?

Haselhorst, L., 1999. Bioremediation of 2,4,6-Trinitrotoluene (TNT) at munitions sites. Restoration and Reclamation Review 4 (7), 9 (Dept. of Horticultural Science, University of Minnesota).

Heap, I., 2014. The International Survey of Herbicide Resistant Weeds. http://www.zoominfo.com/p/Ian-Heap/3455213 (accessed September 2014.). Also available at www.weedscience.org, Accessed September and October 2014.

Huff Post Green, March 5, 2014a. Anti-GMO March on Kauai Draws Thousands. http://www.huffingtonpost.com/2013/09/09/gmo-march-kauai_n_3894816.html (accessed May 2014.).

Huff Post Green, March 5, 2014b. China rejects GMO Corn in Distillers Grains from US; More Rejections Expected. http://www.huffingtonpost.com/2013/12/26/china-rejects-gmo-corn_n_4502719.html (accessed March 2014.).

ISAAA (International Service for the Acquisition of Agri-Biotech Applications, January 14, 2015a. Crop Biotech Update. Cropbiotechupdate=isaaa.org@mail90.atl51.rsgsv.net (accessed January 2015.).

ISAAA (International Service for the Acquisition of Agri-Biotech Applications, February 4, 2015b. Crop Biotech Update. Cropbiotechupdate=isaaa.org@mail89.atl31.mcdlv.net (accessed February 2015.).

James, C., 2013. Global Status of Commercialized Biotech/GM Crops: 2013. Executive Summary. Brief 46. International Service for the Acquisition of Agric-Biotech Applications. C/o IRRI, DAPO Box 7777, Manila,Philippines.

Jensen, D., September/October 2013. Dead End—on Killing the Planet in Order to Save it. Orion. pp. 11–14.

Jeyaratnam, J., 1990. Acute pesticide poisoning: a major global health problem. World Health Statistics Quarterly 43 (3), 139–144.

Kinney, A.J., 1994. Genetic modification of the storage lipids of plants. Current Opinion in Biotechnology 5 (2), 144–151.

Van Montagu, M., October 23, 2013. The Irrational Fear of GM Food. The Wall Street Journal. p. A-15.

Noble, R.K., November/December 2013. Keeping Science in the Right Hands—Policing the New Biological Frontier. Foreign Affairs. pp. 47–53.

Paine, J.A., Shipton, C.A., Chaggar, S., Howells, R.M., Kennedy, M.J., Vernon, G., Wright, S.Y., Hinchliffe, E., Adams, J.L., 2005. Improving the nutritional value of Golden rice through increased pro-vitamin A content. Nature Biotechnology 23 (4), 482–487.

Pauling, L., Corey, R.B., Branson, H.R., 1951. The structure of proteins: two hydrogen-bonded helical configurations of the polypeptide chain. Proceedings of the National Academy of Sciences 37 (4), 205–211.

Reynolds, G., October 26, 2014. Living Better through Science—A Handy Guide to Recent Research about How to Cheat Death (Good Luck). New York Times Magazine. pp. 36–37.

Scorza, R., Callahan, A., Dardick, C., Ravelonandro, M., Polak, J., Malinowski, T., Zagrai, I., Cambra, M., Kamenova, I., 2013. Genetic engineering of *Plum pox virus* resistance: "HoneySweet" plum—from concept to product. Plant Cell, Tissue and Organ Culture 115 (1), 1–12.

Seidler, R.J., September 2014. Pesticide Use on Genetically Engineered Crops. Organic Consumers Assoc. http://www.organicconsumers.org/articles/article_30947.cfm (accessed November 2014.).

Phytoremediation of TNT and RDX. pp. 371–392. In: Singh, S.N., Mishra, S. (Eds.), 2014. Biological Remediation of Explosive Residues. Springer International, Switzerland, p. 405.

Shiva, V., 1993. Monocultures of the Mind: Biodiversity, Biotechnology and Agriculture. Zed Press, New Delhi. 184 pp.

Shiva, V., 1997. Biopiracy: The Plunder of Nature and Knowledge. South End Press, Cambridge, MA. 148 pp.

Shiva, V., 2000. Tomorrow's Biodiversity. Thames and Hudson, London. 144 pp.

Specter, M., August 25, 2014. Seeds of Doubt—An Activists Controversial Crusade against Genetically Modified Crops. The New Yorker. pp. 46–52, 54–57.

Tang, G., Qin, J., Dolnikowski, G.G., Russell, R.M., Grusack, M.A., 2009. Golden rice is an effective source of vitamin A. American Journal of Clinical Nutrition 89, 1776–1783.

Tenner, E., 1996. Why Things Bite Back: Technology and the Revenge of Unintended Consequences. A. A. Knopf, Inc., New York. 346 pp.

Thompson, P.B., 2007. Food Biotechnology in Ethical Perspective, second ed. Springer, New York. 340 pp.

TIME, December 1–8, 2014a. Bananas That Prevent Blindness. p. 92.

TIME, June 30, 2014b. The Enriched "Super Banana" That Could Save Millions of Lives. p. 13.

Washington, H.A., December 31, 2012. Don't Forget Rosalind Franklin. MS Blog Magazine. msmagazine.com/blog/2012/12/31/don't-forget-rosalind-franklin/.

Watson, J.D., 1968. The Double Helix: A Personal Account of the Discovery of the Structure of DNA. A Mentor book–New American Library, NY. 143 pp.

World Health Organization, 2009. Global Database on Vitamin A Deficiency—Global Prevalence of Vitamin A Deficiency in Populations at Risk, 1995–2005. Geneva, Switzerland.

Ye, X., Al-Babili, S., Klöti, A., Zhang, J., Lucca, P., Beyer, P., Potrykus, I., 2000. Engineering the provitamin A (beta-carotene) biosynthetic pathway into (carotenoid-free) rice endosperm. Science 287 (5451), 303–305.

Zimdahl, R.L., 2012. Fundamentals of Weed Science, fourth ed. Academic Press, New York. 648 pp.

Chapter 9

Antibiotics

Chapter Outline

INTRODUCTION

Definition

The term "antibiotic" was coined in 1942 by Selman Waksman, a microbiologist, and his colleagues to describe any substance produced by a microorganism that is antagonistic to the growth of other microorganisms. His definition excluded substances that kill bacteria, but are not produced by microorganisms (e.g., gastric juices, hydrogen peroxide) and synthetic antibacterial compounds such as the sulfonamides. It included penicillin and streptomycin, which are produced by or derived from fungi, bacteria, and other organisms, and can destroy or inhibit the growth of other microorganisms. An antibiotic is therefore one of a large number of chemicals, produced by microorganisms, that have the capacity, even in dilute solution, to inhibit the growth of or destroy bacteria and other microorganisms.

Discovery

It is commonly and incorrectly assumed that penicillin was the first antibiotic. However, a wide variety of treatments that might be considered to act as antibiotic medicines have been used for more than 1000 years. Ancient Egyptian, Persian, and Greek physicians treated patients with compresses and tonics made from a variety of herbs, molds,[1] and organic compounds. Although they were often not

---

1. Mold: growth of minute fungi on vegetable or animal matter, usually as a downy/furry coating. Commonly associated with decay.

Six Chemicals That Changed Agriculture. http://dx.doi.org/10.1016/B978-0-12-800561-3.00009-2

effective and were frequently dangerous, the intent was similar to modern antibiotics. For centuries, medical doctors attempted to cure infections with assorted natural remedies even although they had no knowledge of bacteria, which had not been seen until the 1660s when Anton van Leeuwenhoek, a Dutch tradesman and scientist discovered them. He is considered the Father of Microbiology. He is best known for his work on the improvement of the microscope and his contributions toward the establishment of microbiology. He managed his father's drapery shop and developed an interest in lens making to observe cloth quality. Using his handcrafted microscopes he was the first to observe abundant living things, which he called animalcules, in a sample of pond water.

Conclusive proof of the existence of microorganisms (bacteria) was provided by Louis Pasteur and Robert Koch in the 1860s and 1870s. Pasteur showed that microorganisms grew in broth in a sealed tube, but did not grow if the broth was boiled first. Thus, we have pasteurization. Robert Koch demonstrated that anthrax was caused by a bacterium. His criteria proved the germ theory and Koch's postulates are still used to prove that an infectious agent causes a specific disease.

Some purported cures were based more on superstition and traditional folk remedies than on science. Ointments and potions made from comfrey (*Symphytum officinale* L.) or St John's wort (*Hypericum perforatum* L.) might have had some antibiotic effects. Other treatments, including wine, which was most commonly used as an astringent, sleeping with snakes in a temple, applying salves made from animal dung, and wearing magical talismans have all been rejected by modern medicine. However, it is worth noting that comfrey is still available and advocated in some quarters for treatment of bronchial problems, broken bones, sprains, arthritis, gastric and varicose ulcers, severe burns, acne, and other skin conditions. It is reputed to have bone- and teeth-building properties in children and have value in treating female disorders. St John's wort is grown commercially in the United States for use in herbal medicine, primarily as an antidepressant, psychiatric medication.

In 1888, the antibiotic, pyocyanase, was discovered by the German scientist E. de Freudenreich, who isolated it from a bacterium. The blue pigment in the culture stopped the growth of bacteria. Unfortunately, the disease-fighting antibiotic could not be made into a usable product because it was proven to be toxic and unstable in the body. It was used only as a final effort to save patients who would surely die, and did, even when pyocyanase was administered.

Several scientists were involved with preliminary research before de Freundenreich's work.[2]

In 1640, John Parkington in England recommended using mold for treatment in his book on pharmacology. In 1870, Sir John Scott Burdon-Sanderson observed that culture fluid covered with mold did not produce bacteria. In 1871, Joseph Lister, a British surgeon and pioneer of antiseptic surgery, experimented

---

2. explorable.com/history-of-antibiotics. Accessed March 2014.

with the antibacterial action on human tissue of *Penicillium glaucium*, a mold used in making some types of bleu cheese. In 1874, Sir William Roberts, a physician from Manchester, noted that cultures of *P. glaucium* mold did not display bacterial contamination. John Tyndall explained the antibacterial action of the *Penicillium* fungus to the Royal Society in 1875. Pasteur, a French chemist and microbiologist, built on this discovery, noting that *Bacillus anthracis* (the etiologic (causal) agent of anthrax—a common disease of livestock and, occasionally, of humans) would not grow in the presence of the related fungal mold *Penicillium notatum*. He postulated, in 1877, that bacteria could kill other bacteria. The antibiotic power of *P. glaucium* was independently discovered and tested on animals by Ernest Duchesne, a French physician. He inoculated guinea pigs with a normally lethal dose of typhoid bacilli and showed that the animals would be free of the disease if they were also inoculated with *P. glaucium*. He was 23, unknown, and his work was done 32 years before Fleming's research with the fungus *P. notatum*. The Institut Pasteur did not even acknowledge receipt of Duchesne's 1897 dissertation. It was ignored.[3]

The seminal scientist in the discovery of antibiotics was Sir Alexander Fleming, a Scottish biologist, pharmacologist, and botanist who was born in Lochfield, Scotland, in 1881 (died 1955). He defined new horizons for antibiotic research with his discovery of the enzyme lysozyme in 1923. It destroys the cell walls of bacteria. Unfortunately it, similar to pyocyanase, was too toxic to use in humans. Treatment of human bacterial diseases improved immensely after 1928 after Fleming discovered penicillin produced by *P. notatum*. It became the preferred treatment for bacterial infections such as syphilis, gangrene, and tuberculosis.

In 1927, Fleming was investigating the properties of staphylococci, a common bacterium that lives on the skin and mucous membranes of humans. He returned to his laboratory in September 1928 after having spent August on holiday with his family. Before leaving, he stacked all of his staphylococci cultures on a bench in a corner of his laboratory. When he returned, he noticed that one culture was contaminated with a fungus, and that the colonies of staphylococci surrounding the fungus had been destroyed, whereas other colonies farther away were normal. Fleming was clearly a good observer who saw what he was looking for when it was there, did not see what he was looking for when it was not there, and, most importantly, saw what he was not looking for when it was there. Many would have discarded the culture because they would have assumed it was contaminated or was not consistent with what they were looking for.[4] He was not the first to observe mold killing bacteria on a plate, he was the first to investigate further.

He showed the contaminated culture to his former assistant Merlin Price, who reminded him, "That's how you discovered lysozyme." Fleming grew the mold in a

---

3. http://en.wikipedia.org/wiki/Penicillium_glaucum. Accessed March 2014.
4. Many of the preceding paragraphs on Fleming were derived from http://en.wikipedia.org/wiki/ Alexander_Fleming. Accessed March 2014.

pure culture and found that it produced a substance that killed a number of disease-causing bacteria. He identified the mold as being from the *Penicillium* genus, and therefore named it penicillin. It was the first antibiotic that was safe for people. However, it was not commercially manufactured until 1939, and virtually all production was initially reserved for military use. Civilians only gained access after World War II (Gustafson and Bowen, 1997).[5] Fleming shared the Nobel Prize in Physiology or Medicine in 1945 with Howard Florey (a medical doctor and pathologist) and Ernest Chain (a chemist and pathologist) for their discovery of penicillin and its clinical efficacy. Florey studied the antibacterial properties of penicillin in mice be tested, a crucial step Fleming had not taken. Florey was the first to test penicillin in humans. Chain and Florey reported their work in the *Lancet* in 1940.

In 1939, the American microbiologist René Dubos discovered that the soil microbe *Bacillus brevis* decomposed the capsule of pneumococcus bacteria (a major cause of pneumonia in the late nineteenth century), rendering them harmless. Tyrocidine, the major constituent of the topical antibiotic tyrothricin produced by *B. brevis*, was the first commercially available antibiotic. Its use was limited to treating infections of the skin that resulted from contaminated soil. However, it too was found to be toxic to human blood and reproductive cells.

Sulfonamides (sulfa drugs, released in 1935) were the first antibiotics that were not harmful to the patient. They paved the way for the antibiotic revolution in medicine. The first sulfa drug, Prontosil, was discovered by the German physician and chemist Gerhard Domagk in 1935. He worked in the laboratories of Bayer AG, at that time part of the German chemical trust IG Farben. Sulfa drugs kill bacteria and fungi by interfering with their metabolism. They were the "wonder drugs" before penicillin and are still used today. During the 1940s and 1950s, the aminoglycosides (e.g., streptomycin), chloramphenicol, and tetracycline were discovered.

Classification

Table 9.1 shows some antibiotic chemical groups, the year of introduction, and those used in animals and humans. The general classification scheme in Figure 9.1 is based on bacterial spectrum, route of administration, and activity. Neither classification describes what a particular antibiotic does—what disease(s) it treats, how it works—its mode of action, or how effective it may be compared with others in the same chemical group.

HUMAN USES

Antibiotics are used to treat life-threatening bacterial diseases and other infections caused by protozoa and fungi. They kill or impede development of the causal organisms. Antibiotics enabled treatment of many human diseases that were fatal prior to their development.

---

5. http://inventors.about.com/od/pstartinventions/a/Penicillin.htm. Accessed November 2014.

TABLE 9.1 Year of Introduction and Number of Antibiotics in Some Chemical Groups (Butler and Buss, 2006)

Chemical Group	Year Introduced	Number Available	Example	Used in Animals and Humans
Aminoglycoside	1944	8	Streptomycin	X
Ansamycin	1957	4	Rifamycin	
ß lactam	1941	7	Penicillin	X
Bacterial peptide	1942	10+	Bacitracin	X
Cephalosporin	1945	17	Cephalosporin	
Glycopeptide	1956	3	Vancomycin	
Ionophore	1967	3+	Monensin	Not in humans
Lincosamide	1952	2	Lincomycin	
Lipopeptide	2003	1	Daptomycin	
Macrolide	1952	6	Erythromycin	X
Nitrofuran	1947	2	Nitrofurantoin	
Nitroimidazole	1959	3	Tinidazole	X
Oxazolidinone	2000	1	Linezolid	
Quinolone	1962	10	Nalidixic acid	X
Sulfonamide	1935	5	Sulfapyridine	X
Tetracyclines	1950	6	Chlortetracycline	X
Other—		12		
Against mycobacteria		12		

Note: Antibiotics in the chemical classes Streptogramins, Phenicols, and Pleuromutilin are used exclusively in veterinary medicine. The year of introduction is often an approximation.
http://en.wikipedia.org/wiki/List_of_antibiotics. Accessed March 2014.

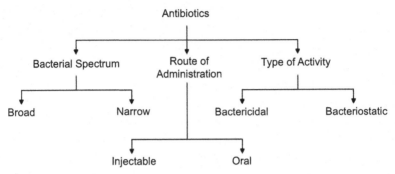

FIGURE 9.1 A classification of antibiotics based on their bacterial spectrum, route of administration, and activity. *Adapted from* Explorable.com.

Macrolides[6] (e.g., erythromycin, tylosin) are bacteriostatic antibiotics that have been among the best-tolerated antibiotics for more than 50 years. Typical treatment is 5–7 days except for some bacterial infections that require medication for 2–3 weeks. There are several well-known advantages to macrolide antibiotics:

- broad antibacterial spectrum
- simple to use
- low incidence of gastrointestinal side effects
- can usually be used by all age groups
- safe during pregnancy.

There are equally well-known disadvantages shared by most antibiotic groups, which vary from mild to severe, depending on the specific antibiotic.

- Some may induce hypersensitivity and cause allergic reactions.
- Possible side effects include vomiting, gastric irritation, and liver complications.
- Inherently resistant portions of any population will survive, become dominant, and will withstand the previously effective antibiotic.
- Previously treatable infectious diseases may be incurable.
- People carrying resistant pathogens may spread them to others.
- Killing large segments, perhaps all, of the normal microbial communities in our body may result in loss of some of the body's beneficial bacteria. The normal balance of existing microflora and microfauna is altered.

There is no question that antibiotics combat a wide variety of infections, it is important to realize that, by definition, they are only effective against bacterial infections. They are not effective treatments for viral infections (e.g., the common cold) or most fungal infections (e.g., ringworm). There are more than 100

---

6. Macrolides, a group of complex drugs (including antibiotics), whose activity stems from the presence of a large macrocyclic lactone ring to which one or more deoxy sugars is attached.

antibiotics available to treat human and animal diseases. Several antibiotics in the chemical groups in Table 9.1 are available under more than one trade name. All work in one of two ways:

1. A bactericidal antibiotic kills bacteria by interfering with the formation of the bacterium's cell wall or its cell contents.
2. A bacteriostatic antibiotic stops bacterial multiplication by interfering with protein production, DNA replication, or other aspects of bacterial cellular metabolism.

Citizens of the world's developed societies live healthier, longer lives because of antibiotics. These drugs are so readily available and effective that we take them for granted and incorrectly assume there must be at least one that will cure almost any malady. When one uses a search engine, it is easy to find lists (e.g., Google, with 62) of antibiotics that treat diseases ranging from acne to bursitis, laryngitis, meningitis, tuberculosis, and vertigo. The human diseases now treated and frequently cured by antibiotics killed or crippled our ancestors in droves. At the beginning of the twentieth century, illnesses caused by infectious agents were among the most common causes of death in North America and worldwide. The advent of antibiotics by mid-century raised hopes that many infectious diseases would be eliminated by the end of the century (Conly and Johnston, 2005).

Alexander Fleming warned that when an antibiotic was used frequently and extensively to treat single or multiple diseases, bacterial resistance would occur. It has! Antibiotic resistance may occur in one of three ways each of which may have human medical consequences:

1. The antibiotic may cause spontaneous bacterial mutation at a low rate, which may be followed by selection of resistant cells in the drug's continuing presence. This is a slow mechanism.
2. Bacteria may acquire genes that code for resistance. These genetic changes alter the bacteria's defensive functions by changing the drugs metabolic target, detoxifying (metabolizing) it, ejecting it, or by altering the metabolic route so it never reaches the point (the disease site) at which it was directed (Witte, 1998).
3. In most cases, antibiotics do not directly cause or create resistance. They select for resistant individuals in a population, just as pesticides do. Table 9.2 shows the development of a resistant population during 7 years (Gressel, 1990). The illustration was prepared to show development of a population of weeds resistant to an herbicide, but the comparison holds. It is highly likely that for the first 3–4 years, the resistant population will not be large enough to have a significant effect or even be noticed; the disease may appear to be cured and the assumption of successful treatment will be logical. Subsequently, the resistant population will dominate and the antibiotic's failure will be obvious. The governing maxim as antibiotics selected for resistant individuals was what doesn't kill a bacterial population makes it stronger and thereby a more pernicious and successful cause of disease.

TABLE 9.2 Development of a Population of Resistant Weeds with Repeated Use of a Single Herbicide (Gressel, 1990)

Year	Susceptible Population	Resistant Population
1	100,000,000	1
2	10,000,000	4
3	1,000,000	16
4	100,000	64
5	10,000	256
6	1000	1024
7	100	4026

Research has shown that the vast majority of antibiotics prescribed for mild sore throats, which are often viral infections, are useless. An editorial (Anonymous, 1996) warned that the "healing power of antimicrobial agents has been threatened by the ever more insistent emergence of antibiotic resistant bacteria." The claim is supported by Keiger (2009). In 2013, the US Centers for Disease Control and Prevention identified 17 antibiotic-resistant microorganisms that caused at least 23,000 deaths in the United States (Allan, 2014). The article strongly recommended that if we don't "curb our use of existing antibiotics and commit to developing new ones, the risks are not just medical but economic" (Allan, 2014). Both warnings echoed Fleming's counsel. The Centers for Disease Control and Prevention estimates that, in the United States, antibiotic resistance now costs $20 billion in excess health-care spending and $35 billion in lost productivity (Allan).

However, the antibiotics pipeline is drying up, whereas resistance to existing drugs is increasing (Braine, 2011). There are several reasons why the pharmaceutical industry's interest in developing new drugs has declined, although there are abundant research projects that could be pursued. The research costs for development of one antibiotic exceed $800 million and the work may take 10 years (Conly and Johnston, 2005), with no guarantee that the product will be therapeutically successful and profitable. There's been much more research focused on products for treatment of chronic disease, in which the potential profits are greater. Although there may be great opportunity to find new therapeutic antibiotics, the profit potential discourages investment.

Antibiotic resistance is not just a challenge to the health of humans and animals in the developed world. More than 200 zoonotic diseases can be transmitted from animals to people. Just six diseases: acute respiratory infections including pneumonia and influenza (2.6 million deaths), AIDS (2.7

million), diarrheal diseases (1.8 million), tuberculosis (0.9 million), malaria (1.1 million), and measles (0.7 million); 2012 data) cause 90% of the deaths of young adults and children in developing countries (WHO, 1999, cited in Meslin et al., 2000, p. 312). The rate at which these diseases have appeared has increased over the past 40 years. Bacterial diseases have a disproportionate effect on poor people in poor countries, where 20% of human sickness and death is due to zoonoses.[7] Worldwide, they are becoming more serious. World Health Organization data from 2001 claimed that communicable diseases, nutritional deficiencies, and maternal and perinatal diseases caused 12.8 million deaths. Diarrheal diseases and lower respiratory infections (including pneumonia) caused 35% of these deaths. Bacterial zoonotic diseases are responsible for 16% of all global deaths and 2/3 of deaths of children younger than age 5. Antibiotic resistance is not the sole cause of these horrific data, but it surely must be recognized as an important contributing factor. Developing countries also need better public health systems and greater public awareness of causes of disease problems and their spread. In addition, the ready availability of and policies regulating use of antibiotics are poorly developed or absent in the world's developing countries (Witte, 1998).

ANIMAL USES

Antibiotics are used to treat and prevent animal diseases, just as they are in humans. The advantages and disadvantages are the same. One difference is that humans decide to use or not use an antibiotic recommended by a physician and obtained with a prescription from a pharmacist. Animal owners decide to use or not use an antibiotic recommended by a veterinarian for their animals. The animals do not choose. A veterinarians prescription is required for some, but not for all.

As noted previously, penicillin was first noticed in 1896, re-discovered by Fleming in 1928, and its anti-bacterial therapeutic efficacy was demonstrated by Chain et al. (1940). It was more effective than all previous treatments and was the first effective antibiotic for treating bovine mastitis. Moore et al. (1946) had found that streptomycin added to the diet of chicks improved their growth. At the time, his work did not lead to further experiments with animals. Several laboratories were searching for an inexpensive source of vitamin B_{12} as a dietary supplement to improve the growth rate of poultry (Jukes, 1977). The B_{12} research included study of its use as an animal feed supplement (Ogle, 2013). That work led to the discovery that the presumably effective vitamin B_{12} supplement was contaminated with the antibiotic aureomycin. It, not B_{12}, was responsible for improved animal growth.

---

7. http://www.ebolaupdate.com/uploads/1/2/0/6/12064646/zoonoses_rapid_response_briefing _final.pdf. Accessed November 2014.

In 1950, American farmers rejoiced at news that scientists had discovered that adding antibiotics to animal feed increased growth and cost less than feed supplements used at the time (Ogle, 2013). Antibiotics are now used in animal and poultry production for four purposes, listed as follows in order of the reason for use.

Therapeutic: treating sick animals
Metaphylaxis: short-term medication to treat a sick animal and prevent disease or infection spread
Prophylactic: to prevent infection during times of risk (transport or weaning)
Growth promotion: to improve feed efficiency and promote faster maturity.

Antibiotics were, and still are, a boon to animal agriculture because they lower production costs, promote faster growth, reduce time to market—which increases growers' profit—and by preventing diseases, they maintain animal's health and prevent losses. The total amount of antibiotics used in United States animal/poultry agriculture is not certain (Mathew et al., 2007). Available estimates are contentious, plagued by conflicting definitions, and there is continuing debate over what should be counted. Most center on 29–30 million pounds of antibiotics used annually for animals/poultry and 7.3–7.9 million pounds used for human disease control. Thus, 70–80% of all antibiotics are used annually in the United States by the animal industry (see footnotes 8–14) (Carolan, 2011). Mathew et al. (2007, p.117) reported that 40 antimicrobial products were approved for use in livestock, of which 14 were also used in human medicine. Table 9.1 identifies eight chemical classes that are used in humans and animals. US Food and Drug Administration (FDA) data from 2009 identify that of 18 chemical classes of antibiotics, 6 of the classes and 11 of 42 antibiotics are used only in animals. The same data for 2011 identify the same 18 chemical classes and 9 of 42 used only in animals (FDA, 2014).

Comparisons between human and animal antibiotic use involve at least five often neglected differences in the use of antibiotics in human and veterinary

---

8. http://www.scientificamerican.com/article/antibiotic-in-food-animals/. Accessed April 2014.

9. http://www.fda.gov/AnimalVeterinary/SafetyHealth/AntimicrobialResistance/NationalAntimicrobial ResistanceMonitoringSystem/ucm095684.htm. Accessed April 2014.

10. http://www.mindfully.org/Health/Antibiotics-Healthy-Livestock-UCS.htm. Accessed April 2014.

11. http://www.motherjones.com/tom-philpott/2013/02/meat-industry-still-gorging-antibiotics. Accessed April 2014.

12. http://mrsatopic.com/2012/07/antibiotic-contamination-of-water-what-are-the-long-term-effects-on-us/. Accessed April 2014.

13. http://www.scientificamerican.com/article/most-us-antibiotics-fed-t/. Accessed April 2014.

14. http://www.wired.com/2010/12/news-break-fda-estimate-us-livestock-get-29-million-pounds-of-antibiotics-per-year/. Accessed April 2014.

medicine that should be considered in developing the case for or against animal use and its relationship to antibiotic resistance. Therefore

- the number of humans in the population compared with the number of animals in each of the many veterinary populations.
- Differences in physical characteristics of humans and various animal species (e.g., weight).
- Antibiotic use in humans can be for the treatment or prevention of an infection, whereas animal use may include treatment, control, prevention, and growth promotion. Available data do not permit distinguishing between these different uses. The majority of antibiotics used in animal feed are approved for both therapeutic and production purposes. Therefore, the route of use cannot be used as a proxy that indicates the primary purpose of use and may obscure a secondary purpose.
- Dosages for different antibacterial drugs differ between humans and animals as they do between young and adult humans.
- Duration and dose of antibiotic drugs varies between animals and humans.

For these reasons, it is difficult to draw definite conclusions from direct comparisons between the quantity of antibacterial drugs used in humans and in animals.[15]

The clear justification to assure the health of animals and poultry, whose meat, milk, and eggs are destined for the grocery store near you, is based on two arguments. The first is that it is more economical to prevent a disease than to treat it. That is, use of prophylactic antibiotics is beneficial. The second argument is based on the overall economic benefits of prevention. Both reflect the quite logical premise that animal agriculture, as is true for all agriculture, is designed to produce food at a profit for the producer.

THE CONTROVERSY

The arguments in favor of prophylactic antibiotic use are countered by four claims:

- Heavy, indiscriminate, continuous use of antibiotics in poultry, cattle/swine (meat), and dairy product production may result in the transfer of residual antibiotics to human food.
- Antibiotic resistant bacteria can be transferred from humans to animals.
- Perhaps of greatest concern is that this will lead to ineffectiveness of antibiotics for treatment of human diseases.
- Resistant genes could also be transferred to another pathogenic species thus spreading a health hazard.

---

15. http://www.fda.gov/Drugs/DrugSafety/InformationbyDrugClass/ucm261160.htm. Accessed November 2014.

Quite simply, the controversy is first about whether or not "antibiotic use in animal husbandry is a driving force for the development of antibiotic resistance in certain pathogenic bacterial species" (Witte, 1998).[16] Second, if the answer to the first question is yes, does that have important, negative human health implications? These have been the major public health questions for at least 40 years, but in recent years they have become more important and controversial. Scientific research is important to resolution of the questions. However, the questions are not strictly scientific. They are value issues. Science produces rational truths that can be defined mathematically, are publicly verifiable, literal, definitive, falsifiable, and precise. In contrast, experiential truth (values) uses, and depends on language, which is frequently vague, imprecise, nonliteral, symbolic, descriptive, and highly subjective. It deals with what is most important, what has the highest value (Zimdahl, 2012, p. 3). Scientific evidence is essential to the discussion, but experiential truth may become the most important realm for resolution of the controversy described above.

Concentrated Animal Feeding Operations

A major agricultural development—indeed a transformation of animal and consequently all of agriculture enabled by antibiotics—is the development and now widespread presence of concentrated animal feeding operations (CAFOs), which:

1. confine animals for more than 45 days during their lifetime,
2. in an area that does not produce vegetation, and
3. meets certain size thresholds (see Table 9.3).[17]

CAFOs transformed agriculture in the last decades of the twentieth century and continue to do so. In large measure, they have created greater public awareness and concern about the two questions posed previously. The transformation has contributed to the slow, inevitable disappearance of livestock production on small family farms and the gradual demise of the communities and small businesses those farms created and supported. Many dismiss concern about these losses as the price of progress. The concern exemplifies a value as opposed to a scientific issue.

Production of meat, milk, and eggs in CAFOs concentrates large numbers of livestock or poultry in relatively small, confined places, and substitutes external management, structures, and equipment (for feeding, temperature control, and manure management) for land and a farmer's skill and labor. It

---

16. www.scientificamerican.com/article/antibiotic-in-food-animals/. Accessed April 2014.
17. http://en.wikipedia.org/wiki/Concentrated_Animal_Feeding_Operations. Accessed May 2014.

TABLE 9.3 US Environmental Protection Agency Definitions of Concentrated Animal Feeding Operation Sizes

Livestock	Operational Animal Capacity		
	Large	Medium	Small
Cattle or cow/calf pairs	1000+	300–999	Less than 300
Mature dairy cattle	700+	200–699	Less than 200
Sheep or lambs	10,000+	3000–9999	Less than 3000
Turkeys	55,000+	16,500–54,999	Less than 16,500
Laying Hens or Broilers			
Liquid manure system	30,000+	9,000–29,999	Less than 9000
Other Chickens			
Not a liquid manure system	125,000+	37,500–124,999	Less than 37,500
Laying Hens			
Not a liquid manure system	82,000+	25,000–81,999	Less than 25,000

has been and continues to be a process that values quantity over quality of product and neglects its externalized[18] effects. These effects include those on

- People: poor air quality (noxious odors), public health, product quality, small farmers;
- The environment: water, air quality, manure disposal;
- The economy: loss of small businesses, small farms, and rural communities; and
- Concern about animal welfare.

Carolan (2011) asserts that "evolution has not prepared cattle for the 3 to 6 months they spend in a feedlot." The corn-rich diet can lead to acidosis, which can kill an animal or make it sick and thus "off its feed." The claim is

---

18. An externality is a cost that is not reflected in price. In the accounting sense, it is a cost that a decision-maker does not have to bear or a benefit that cannot be captured. It is a secondary cost or benefit that does not affect the decision-maker. It can also be viewed as a good or service whose price does not reflect the true social cost of its consumption.

that modern, intensive animal agriculture requires feeding antibiotics in proper therapeutic, metaphylaxic, or prophylactic doses to maintain animal welfare by preventing diseases. The contrary claim is that the present system of large-scale, confined animal agriculture is inherently harmful to animal welfare and that a significant percentage of antibiotics fed to livestock are used solely for growth promotion.

There is no question that much (not all, e.g., Western cattle ranches) animal agriculture is now done in large CAFOs. They are economically rational, efficient agricultural enterprises, in which animals produce large, indeed huge, amounts of manure and liquid waste, which is concentrated as the production facility is. The US Department of Agriculture estimates that US livestock and poultry produce 335 million tons of manure/year (40 times as much as humans produce), which is one way resistant pathogens get into the environment (Keiger, 2009). When possible, CAFO managers use it for fertilizer. Before spreading it on fields, manure is collected in sheds and manure lagoons. Disposing of it is a problem because of its volume; reservoir of antibiotic resistant bacteria; toxicity from its unavoidable, enormous content (up to 10 million/liter of manure) of *Escherichia coli* infective bacteria; and high concentrations of nitrogen, other nutrients, and heavy metals (e.g., As, Cd, Cu, Zn). Putting manure on crop fields is a good thing because it restores organic material to soil. However, it is not valuable, and hauling it long distances to apply to distant fields is economically unwise. Pollution is an undesirable, perhaps inevitable, result and an example of externalization of costs.

There are roughly 257,000 CAFOs in the United States, of which about 15,500 meet the narrow criteria for CAFOs (see the three criteria discussed previously). The US Environmental Protection Agency has defined three capacity categories of CAFOs.[19] (Table 9.3) The number of animals for each category depends on species, operational capacity, and the system for manure disposal.

It is important to note that the numbers shown for the large category in Table 9.3 are a minimum size for the category. Factory farms with as many as 30,000 broiler chickens in one building or 100,000 laying hens in another are not uncommon. There are dairy herds with as many as 30,000 milking cows, swine CAFOs with 100,000 pigs, and cattle CAFOs with 40,000 head.

CAFOs dominate livestock and poultry production in the United States and some other countries (e.g., Brazil, China, India, Indonesia, several European Union countries, Thailand, Vietnam) (Leonard, 2014, p. 306). In 2006, Nierenberg reported that 74% of the world's poultry, 43% of beef, and 68% of eggs were produced in CAFOS. In the United States, in 2000, four companies (Tyson, Cargill, JBS, and Smithfield) produced 85% of beef cattle, 73% of sheep, 65% of pork, and 50% of poultry (Leonard) primarily in CAFOs. In 1966, 1 million

---

19. http://en.wikipedia.org/wiki/Concentrated_Animal_Feeding_Operation. Accessed November 2014.

farms raised 57 million pigs; by 2001, it only took 80,000 (12%) to raise the same number of pigs.[20] In 2011, 69,100 US pig farms (farms with more than 2000 pigs) were 87% of the total. Most US beef, chicken, pork, milk, and eggs were produced in CAFOs by the 1970s and 1980s. Their dominance in livestock and poultry production and their market share is steadily increasing (Leonard, 2014).

EFFECTS ON AGRICULTURE

Citizens of the world's developed nations are concerned about several aspects of modern agriculture including, but not limited to pesticides in soil, water, and food; corporate agriculture; biotechnology/genetic modification; animal treatment; pollution from animal factory wastes; and possible negative effects on humans of widespread use of antibiotics in animal agriculture.

If these are just fringe concerns of a small, vocal minority of the population, they can be, and are, largely ignored by what one might call the agricultural establishment. However, if they are concerns of the general society, the agricultural enterprise on which we all depend has an ethical dilemma.

The National Institute for Animal Agriculture commissioned a market research study in 2013 and acknowledged their surprise at the scope of consumer awareness and concern about animal production practices. Consumers were particularly concerned about the use of antibiotics and hormones in meat products. The science is clear that antibiotic overuse in agriculture is dangerous, but regulation by the US FDA has changed little since the late 1970s.[21] However, in November 2014, four US senators asked the FDA to improve data collection on antibiotic use and resistance in animal agriculture.

The American Medical Association and the American Public health Association and other health and environmental advocates cite dozens of studies that support their contention that feeding medically important antibiotics to livestock generates drug resistant pathogens that pose a significant threat to human health. The American Veterinary Medical Association, the drug, and meat industries counter with studies that indicate that "the human health risk posed by current livestock antibiotic practices is extremely small" (Downing, 2011a). These competing estimates have played a major role in the public relations battle over the issue. The accused, the latter group mentioned previously, tend to minimize the effects of animal antibiotic use. They assert that "per unit biomass, human and companion antibiotic use is 10 times greater than food animal use." A claim discredited by 2011 FDA data that show livestock use of 28.9 million pounds in 2009 and human use of 7.3 million pounds.

---

20. http://www.happy-mothering.com/07/household/what-is-a-cafo/. Accessed November 2014.
21. www.wired.com/wiredscience/2013/08/fda-commish-change/. Accessed April 2014.

More recently, the *Economist* (2014) reported that antibiotic resistance is costing human lives. In an increasing number of countries most cases of urinary tract infection are resistant to the standard, short-course antibiotic treatment. "At least 2 million Americans are thought to suffer antibiotic-resistant infections each year, leading to some 23,000 deaths" and many health complications from other illnesses. Nationwide, there were 900,000 cases of antibiotic-resistant infections in 2000. The total societal costs were estimated to be $35 billion. It is also pointed out that the rate of antibiotic-resistant infections has more than doubled since 2000.[22] The *Economist* article posits that the cause is overuse by humans taking them for illnesses where they are known to be ineffective and "by farmers feeding them to animals to promote growth." In the same month, TIME (2014), citing the same data, claimed that we are losing the "man versus microbe battle." The situation is complicated by the fact that, as mentioned previously, discovery of new drugs is diminishing. Since 2009, only two new antibiotics have been released in the United States, and the number of new antibiotics annually approved for marketing continues to decline.[23]

At least 2 million people are sickened and an estimated 99,000 die every year from hospital-acquired infections, the majority of which result from the lack of effectiveness of available antibiotics against resistant bacteria. It is unknown how many of these illnesses and deaths result from agricultural uses of antibiotics (Gardner, 2012), but the prevailing assumption is that many do (Anonymous, 2014; Gardner, 2012; Keiger, 2009).

The debate about antibiotic use in animal agriculture goes on (Downing, 2011b). One must ask: Why? There is enough scientific evidence to affirm a link between nontherapeutic antibiotic use in livestock of all kinds, especially when raised in CAFOs, and subsequent negative effects on human health resulting from antibiotic resistance. The evidence is sufficient to satisfy those who demand science-based decisions to resolve the economic, environmental, animal welfare, human health claims, and address competing value claims (Keiger, 2009). Rollin's (2001) assertion is correct: "the evidence is sufficient to curtail non-therapeutic antibiotic use in livestock." "Such curtailment will not harm consumers significantly, won't harm developing nations' evolving agriculture, and could produce hitherto unnoticed benefits, namely restoring the possibility of a more husbandry-based, sustainable agriculture to replace the high-tech agriculture that has hurt animals, the environment, small farms, and sustainability" (Rollin, 2001).

However, the animal, veterinary medicine, drug, and meat industries are not satisfied that the science-based evidence, they also advocate, is sufficient for curtailing nontherapeutic use. Antibiotics have dramatically changed the way

---

22. http://www.prnewswire.com/news-releases/antibiotic-resistant-infections-cost-the-us-healthcare-system-in-excess-of-20-billion-annually-64727562.html. Accessed November 2014.
23. http://en.wikipedia.org/wiki/Infectious_Diseases_Society_of_America. Accessed November 2014.

our meat, milk, and eggs are produced. Because the grocery store is always full of what we want, the cost is low (about 10–11% of income[24]), and public pressure for action is limited to absent. Most people are unaware of the problem. As Leonard (2014, p. 233) pointed out: "nobody seemed to care." The debate goes on within and outside the agricultural community. There is no doubt that it is and will continue to affect the US livestock industry. If public opposition to continued nontherapeutic antibiotic use in agriculture prevails it will cause major changes in the livestock industry.

REFERENCES

Allan, N., March 2014. We're running out of antibiotics. The Atlantic 34.

Anonymous, August 4, 1966. Infectious drug resistance (editorial). New England Journal of Medicine 275 (5), 277. Also see: Http://www.scientificamerican.com/article/antibiotic-in-food-animals/ (accessed April 2014).

Anonymous, January 18, 2014. On the zoonose. Economist 62.

Braine, T., 2011. Race against time to develop new antibiotics. Bulletin of the World Health Organization 89, 88–89. http://dx.doi.org/10.2471/BLT.11.030211.

Butler, M.S., Buss, A.D., 2006. Natural products—the future scaffolds for novel antibiotics? Biochemical Pharmacology 71 (7), 919–929.

Carolan, M., 2011. The Real Cost of Cheap Food. Earthscan. Abingdon, Oxon, UK. 272 pp.

Chain, E., Florey, H.W., Gardner, A.D., Heatley, N.G., Jennings, M.A., Orr-Ewing, J., Sanders, A.G., 1940. Penicillin as a chemotherapeutic agent. The Lancet 236 (6), 226–228.

Conly, J.M., Johnston, B.L., 2005. Where are all the new antibiotics? The new antibiotic Paradox. Canadian Journal of infectious diseases and medical microbiology 16 (3), 159–160.

Downing, J., May 26, 2011a. Antibiotics: precaution vs proof. The VIN News Service. Also see: http://news.vin/VINNews.aspx?articleId=18679 (accessed April 2014).

Downing, J., May 25, 2011b. FDA: food-animal antibiotic consumption dwarfs human medical use. The VIN News Service. Also see: http://news.vin/VINNews.aspx?articleId=18659 (accessed April 2014).

Economist. 2014. Antibiotic resistance – the drugs don't work. May 3. p. 54.

FDA. U.S. Food and Drug Administration, 2014. 2011 Summary Report on Antimicrobials Sold or Distributed for Use in Food-producing Animals. Washington, DC.

Gardner, H., April 11, 2012. New prescription requirement will cut use of antibiotics in livestock, F.D.A. Says. New York Times. A19.

Gressel, J., 1990. Synergizing herbicides. Review Weed Science 5, 49–82.

Gustafson, R.H., Bowen, R.E., 1997. Antibiotic use in animal agriculture. Journal of Applied Microbiology 83, 531–541.

Jukes, T.J., 1977. The history of antibiotic growth effect. Federation Proceedings 37, 2514–2518.

Keiger, D., 2009. Farmacology. John Hopkins Magazine 61 (3), 24–31.

Leonard, C., 2014. The Meat Racket – the Secret Takeover of America's Food Business. Simon & Schuster, New York. 370 pp.

---

24. http://www.huffingtonpost.com/2012/08/02/food-spending_n_1734946.html. Accessed September 2014.

Mathew, A.G., Cissell, R., Liamthong, S., 2007. Antibiotic resistance in bacteria associated with food animals: a United States perspective of livestock production. Foodborne Pathogens in Disease 4 (2), 115–133.

Meslin, F.X., Stöhr, S., Heymann, D., 2000. Public health implications of emerging zoonoses. Review Science Technical Office International Epiz., [Scientific and Technical Review of the Office International des Epizooties (Paris)] 19 (1), 310–317.

Moore, P.R., Evanson, A., Luckey, T.D., McCoy, E., Elvehjen, C.A., Hart, E.B., 1946. Use of sulfasuxidnine, streptothricin, and streptomycin in nutritional studies with the chick. Journal of Biological Chemistry, 165, 437–441.

Nierenberg, D., 2006. Rethinking the Global Meat Industry. State of the World 2006. Worldwatch Institute, Washnigton, DC. pp 24–40.

Ogle, M., 2013. Riots, Rage, and Resistance: A Brief History of How Antibiotics Arrived on the Farm. http://blogs.scientificamerican.com/guest-blog/2013/09/03 (accessed April 2014).

Rollin, B., 2001. Ethics, science, and antimicrobial resistance. Journal of Agricultural and Environmental Ethics 14 (1), 29–37.

TIME, May 19, 2014. Man vs Microbe – We're Losing the Battle against Bacteria. Can We Win the War? p 20.

WHO (World Health Organization), 1999. Removing Obstacles to Healthy Development. WHO Report on Infectious Diseases. Document No. WHO/CDS/99.1. WHO, Geneva. 68 pp.

Witte, W., 1998. Medical consequences of antibiotic use in agriculture. Science 179, 996–997.

Zimdahl, R.L., 2012. Agriculture's Ethical Horizon. Elsevier, London, UK. 274 pp.

Chapter 10

Conclusion

Science, my lad, is made up of mistakes,
but they are mistakes it is useful to make,
because they lead little by little to the truth.

Verne (2009, p. 157).

Chapter Outline

INTRODUCTION

Chemistry is a science that seeks to know what substances are composed of, how their properties relate to their composition, and how one substance interacts with another. Answers to these questions are of interest to chemists and to many others. Farmers who grow crops need to be aware of the chemistry of life processes, the chemical behavior of production materials (e.g., fertilizers, pesticides), the complex structure and content of soils, and the interactions of soil and water on growing plants. These bits of chemical information are essential to those who produce our food even though the details of the foundational chemistry may be unknown.

Chemists seek to understand and explain the complexity of the many chemical substances that are essential to agriculture, indeed to life. Similarly, medical doctors and others who protect our health need to be familiar with numerous chemical reactions that govern life and how these can be affected or altered by complex medicinal chemicals. Engineers need to know the properties and behavior of materials under varying conditions. Although it is not obvious to many people, chemistry dominates our lives. Its dominance is exemplified by the phrase "Better living through chemistry," a variant of a DuPont advertising slogan, "Better things for better living…through chemistry" adopted in 1935 and retained until 1982 when "through chemistry" was dropped. Since 1999, their advertising slogan has been "The miracles of science."

The scientific advances that have changed agriculture have involved a host of scientists in public institutions and industrial companies. Beginning

Six Chemicals That Changed Agriculture. http://dx.doi.org/10.1016/B978-0-12-800561-3.00010-9

in 1901, Nobel prizes have been awarded in five specific areas of scientific endeavor: physics, chemistry, physiology or medicine, literature, and peace. The economics prize was first awarded in 1968. There is no Nobel Prize in agriculture. However, it is worth noting that this book includes 15 distinguished recipients (seven in chemistry) of the 889 Nobel laureates awarded since 1901. The 15 Nobel laureates are part of the story of how chemistry changed agriculture (Table 10.1).

Scientific advances in chemistry and the life sciences have alleviated some of the most difficult challenges of producing sufficient food for a growing population, protecting human health, and enabling environmental stewardship. Although insects, diseases, and weeds, which decrease crop yield, have not been eliminated, the production risk has been significantly reduced. Science and the resulting technology have made it possible to grow crops and livestock using insecticides, fungicides, herbicides, fertilizer, antibiotics, and, more recently recombinant DNA technology to increase crop yield.

THE CHALLENGE

Agriculture's traditional and continuing challenge to feed a growing world population has not been achieved. In 2013, the United Nations Food and Agriculture Organization estimated that there were 1.02 billion undernourished people, primarily in Asia, which has 16% of the world's arable land and in November 2014 had 60% of the world's 7.27 billion people.

China, with 1.4 million people in 2014, had 32% of Asia's population, growing at only 0.5%/year (a declining population), but only 15% arable land. In contrast, India has 29.4% of Asia's people, growing at 1.3%/year (a growing population), 55% arable land, and will soon be the most populous of Asia's countries.

The principal cause of malnutrition is not simply agriculture's failure, although there is much room for improvement in developing countries; it is poverty, which many attribute to the operation of the world's economic and political systems. Malnutrition affects approximately one-third of the children in developing countries; more than 70% of Asia's children are food insecure. The sad truth is that enough food is produced by the world's farmers to provide approximately 2.5 kg/person/day containing 2500 cal/kg, which is equivalent to 6.25 kilocalories (kcal)/person/day for all of the world's 7.27 billion people; more than enough for an adequate diet. What's the problem? More accurately, what are the problems? Eight, not all equally important, factors are involved.

1. Poverty. People don't have money to buy food, land to produce food, or, if they have land, they do not have adequate, appropriate agricultural technology or adapted plant varieties.
2. Lack of investment in agriculture.
3. Natural disasters (e.g., drought, tropical storms, flooding).

TABLE 10.1 Nobel Laureates Who Made Major Contributions to the Transformation of Agriculture

Name	Year	Prize Awarded in	Agricultural Contribution	Chapter
Paul Ehrlich	1908	Physiology or medicine	Antibiotics	9
Albrecht Kossel	1910	Physiology or medicine	Recombinant DNA	8
Fritz Haber	1918	Chemistry	Nitrogen	4
Carl Bosch	1931	Chemistry	Nitrogen	4
Ernest Chain	1945	Physiology or medicine	Antibiotics	9
Alexander Fleming	1945	Physiology or medicine	Antibiotics	9
Howard Florey	1945	Physiology or medicine	Antibiotics	9
P.H. Müller	1948	Physiology or medicine	DDT	4
Selman Waksman	1952	Physiology or medicine	Antibiotics	9
Linus Pauling	1954	Chemistry	Recombinant DNA	8
Francis Crick	1962	Chemistry	Recombinant DNA	8
James Watson	1962	Chemistry	Recombinant DNA	8
Maurice Wilkins	1962	Chemistry	Recombinant DNA	8
Robert Woodward	1965	Chemistry	Quinine	7
Werner Arber	1978	Physiology or medicine	Recombinant DNA	8

4. Refugees from wars and terrorism.
5. Food wasted in developed countries (see Gunders, 2012 and Fox, 2013 in Chapter 4).
6. Much of what is produced is lost due to poor storage facilities, rodents, and infestations of disease organisms and insects.

7. The world's agricultural system produces enough kilocalories to feed 10 billion persons. It is not sufficient to feed the present population, because too many carbohydrates, and not enough protein and fat, are produced.[1]

8. Distribution. There are two aspects to the distribution problem: (1) the lack of adequate receiving facilities for food and poor roads and equipment to distribute food and (2) a decrease in funds allocated to food aid by developed countries.

These eight things are all important aspects of malnutrition/undernourishment. Another important aspect is there is no human right to food. We all may be or be challenged to be our neighbor's keeper, but for most of us our neighbor is just that, the person next door or a close friend. Singer (2009) strongly suggests we need to greatly expand our concept of who is our neighbor. If we turn our backs on a fifth of the world's people, who go to sleep hungry every night, we will have decided that there is no right to food. We recognize and fulfill our obligation to our neighbor(s), but assume we have no obligation to feed others regardless of who they are or where they are. Singer challenges us to become part of the solution to the agricultural challenge: to feed the world.

Obviously, these are not problems chemists can solve, although they will surely continue to contribute to agricultural technology that will enable the projected 3000 kcal/person/day to be produced and, one hopes, made available to all in some future time.

Over all of these factors looms the continuing growth of the world's human population—projected to be 8.3–10.9 billion by 2050. If growth continues at its current rate (1.1%/year), the number of people could be 9.3 (number of children/woman decreases) or 12.6 billion by 2100[2] (number of children/woman is constant or increases) (Kunzig, 2011, 2014). In addition, the world's population is moving away from a pyramidal shape, in which the majority is younger than 15, to a rectangle, in which the number in each 5-year age cohort will be equal. By 2050, the number of people older than age 50 will be about the same as those younger than 15. Human societies will no longer be dominated by the young, a dramatic change in societal structure.

THE CHANGES: THE TRANSFORMATION OF AGRICULTURE

Farmers, agricultural companies, and research institutions, especially in developed countries, are moving toward sustainable use of the Earth's resources, conservation of water, and reducing, if not eliminating, soil erosion while continuing to increase production. Genetically modified crops, the most rapidly adopted agricultural technology ever developed (see Chapter 8), are the latest

---

1. https://hungermath.wordpress.com/2012/10/25/how-much-food-does-the-world-produce-in-one-year/. Accessed November 2014.
2. http://en.wikipedia.org/wiki/World_population. Accessed September 2014.

contribution of chemistry to the continuing transformation of agriculture that enables further progress toward, but alone cannot achieve, the ultimate goal of feeding all.

Precision agriculture may be the next major agricultural change. It is a system that gives farmers precise information about what to plant, what pesticides are required, and how much fertilizer to use on particular soils and even on different soils in a field. The concept, conceived in the early 1980s, involves farm management based on observing, measuring, and responding to inter- and intrafield spatial and temporal variability in crop production. The ultimate goal of precision agriculture is to create farm/crop/livestock management systems that optimize returns, minimize inputs, preserve resources, and protect the environment. In a very real sense, it may herald the ecological intensification of agriculture (Hubert et al., 2014). It has been described as a kind of agronomy, crop, and livestock production that cooperates with nature and ecological processes rather than fighting against them. It will require less fossil fuel through new, less-intensive, tillage techniques. It will mobilize soil microorganisms to build organic matter, control pests by integrated pest management methods (i.e., biological control), and develop specific, longer crop rotations using precise field organization. Ecological intensification will build on traditional knowledge, while using and participating in development of scientific and technological innovations, including genetic modification (Hubert et al.). Ecological knowledge and concerns will become a central part of agriculture. The responsibility of the farmer and the scientist will become cooperation with and learning from the natural world rather than continuing the presently dominant, usually unspoken Biblical injunction[3] to subdue and have dominion over the earth. Nanotechnology may also make, presently unknown, but significant contributions to production agriculture (David and Thompson, 2008).

Advocates of precision agriculture are confident it will boost yields everywhere. It, similar to genetic modification, will be controversial. The major questions and debates will focus on, who owns the information and who benefits most from the technology. If farmers are required to assume a disproportionate share of the risk while the developers of the technology receive most of the benefits, there will be problems (Economist, 2014a). Those engaged in agriculture celebrate their achievements and see a bright future aided by new technological developments. It is equally clear that not all segments of society celebrate agriculture's productive achievements. Many see agriculture's accomplishments as troublesome or even fearful. As one considers the changes chemistry has contributed to the future of agriculture—the essential human activity.

---

3. "…and God said to them. Be fruitful and multiply, and fill the earth, and subdue it…" (New American Standard Bible). The King James Bible version is "and let them have dominion…"; Genesis 1:26, 28.

Dickens' (1859), opening paragraph in *The Tale of Two Cities* addresses both sides of the controversy while satisfying neither.

> It was the best of times,
> it was the worst of times,
> it was the epoch of wisdom,
> it was the age of foolishness.

THE FUTURE: A BRIEF COMMENT

Avery pointed out in 1997 that without the crop yield increases that occurred since 1960, the world would have required an additional 10–12 million square miles (roughly the land area of the United States, the European Union countries, and Brazil combined) for agriculture, bound by 1950s technology, to achieve 1997 levels of food production. Avery claimed that modern high-yield agriculture is not one of the world's problems, but rather the solution to providing sufficient food for all, sufficient land for wildlife, and protecting the environment. His claim is still valid and the need for further changes, many from chemical research, is unconditionally supported by the fact that all of the world's land that is capable of supporting agriculture is now used and, barring a worldwide disaster, the human population will continue to increase for several decades. There is no land for agriculture to expand. That is not news to international organizations, most countries, nongovernmental organizations, companies engaged in agriculture, faculty of Colleges of Agriculture, and farmers. It may be a revelation to the vast majority of people who don't farm. After all, how can there be a problem, the grocery store is always full.

Le Couteur and Burreson (2003, p. 351) correctly suggest that historical events almost always have more than one cause. Each of the chemicals and chemical reactions included in this book is a fundamental part of the changes in US agriculture that contributed to its enormous productivity—the envy of the world. Development of the agricultural potential of each of the chemicals herein can properly be regarded as a historical event, but not one of them is the sole cause of modern agriculture's productivity. Combined, there should be no debate that they have contributed significantly to the productivity of modern agriculture. Together they enabled farmers to achieve a greater degree of long-range predictive control over production and our food supply.

The agricultural revolution gave rise to early civilizations. However, science and technological advances are not always exclusively good. Diamond (1987) regarded settled agriculture as "the worst mistake in the history of the human race." He does not discount the contributions of chemistry and science to increasing agricultural production. He asks: why did almost all our hunter-gather ancestors adopt settled agriculture? Those who espouse what Diamond calls the "progressivist perspective" claim it is because settled agriculture was a more efficient way to get more food for less work. However, no one has proved that the progressivist interpretation is correct. It has been accepted as fact and not questioned. Diamond argues that settled agriculture was bad for human health, starvation became more possible because of dependence on a limited number of crops, and because people settled and carried on trade with other settled societies, the spread of parasites and infectious diseases increased. Settled agriculture—farming, in his view—led to deeper class divisions in society and inequality between the sexes. In his conclusion, Diamond does not ask, but strongly implies, that even a casual examination of today's societies will support his pessimistic claims.

Suppe (1987) offers a different analysis of the progressivist view. "Scientific agriculture's rapidly rising production released millions from farming." They found employment readily in the industrial revolution of developing companies that produced the machines and the chemical technology that enabled rising production by fewer farmers (see Chapter 2, Table 2.4). The industrial and agricultural revolutions of the Western world developed together and complemented each other. Agricultures increasing productive efficiency required less labor, which found employment in industries, some of which created the technology that enabled agriculture's productivity. Together, in a very real sense, they created modern societies.

Modern, scientific agriculture has successfully reduced the cost of food relative to income. In 1930, 20% of a family's income was spent on food, most for home consumption. In 2012, it had dropped to 5.7% for home consumption and 4.3% for food away from home. Suppe goes on to mention that an important negative cost of modern technology, which increased production efficiency, drove farmers into bankruptcy, and out of business. Because of its human costs, Suppe disputes, as Diamond does, the progressivist view that scientific agriculture has been only beneficial. The advocates of the benefits of modern agriculture err when they ignore or fail to consider the unanticipated consequences and externalities of modern agriculture, including its chemical technology, because these, "more often than not, add additional layers of problems to existing problems" (Kunstler, 2012). They are frequently dismissed as the price of progress.

The optimistic future described by John Maynard Keynes (1883–1946), the premier economist of the twentieth century, whose ideas fundamentally affected the theory and practice of modern macroeconomics and informed the economic policies of governments, was not achieved. Keynes (1930) predicted that by 2028, living standards in Europe and the United States would be improved so that no one would need to worry about making money. He thought his grandchildren would work about 3 h a day and even that would be more than was necessary to achieve a good life. The nineteenth century's technological innovations of electricity, fuel, steel, rubber, machinery, chemicals, and the methods of mass production, in Keynes view, made further economic growth inevitable, but more work would not be required. The agricultural sector's enormous increases in production efficiency, farm size, and its reduction in the need for labor, fulfilled Keynes predictions.

However, his prediction for the general economy was too optimistic. He misread human nature (Kolbert, 2014). He assumed that people work to earn enough to buy what they need. He concluded that rational people will therefore choose to work less and have more leisure time. What happened was the opposite. What Keynes missed is that human wants seem to expand exponentially, resulting in more work and less leisure time.

Chemical information is essential to those who produce our food, even though the details of the foundational chemistry may be unknown. The scientific advances that have changed agriculture have involved a host of scientists, including 15 Nobel laureates. The three naturally occurring, two synthetic, and the two chemical discoveries in this book each contributed new, still essential, agricultural technology that made it possible increase yields of crops and livestock. Nevertheless, the challenge of feeding all has not been achieved. Research by agricultural and chemical scientists will be fundamental to developing new technological innovations that help all involved in agriculture meet the challenge of feeding all.

Cooperative research between agricultural scientists and chemists will be fundamental to developing presently experimental and other presently undreamed of technological innovations that will meet the challenge of feeding all. A few examples of the possible results of cooperative research include the following.

- Batteries that are smaller, more powerful, and longer lasting than ever imagined will affect agriculture. Laptop computers, cell phones, and hybrid microsystems will be able to continuously monitor cropped fields to indicate when irrigation, additional fertility (even in parts of a field), or weed or insect control is needed.
- New batteries will enable field monitoring by drones. In France, a drone using infrared, near-infrared, and visible wavelengths records soil moisture, chlorophyll content of leaves, and plant biomass. The data are analyzed remotely, downloaded to a computer, and then uploaded to a tractor-mounted global positioning system, which will forecast irrigation need and apply optimal amounts of fertilizer and pesticides to segments of a cropped field. Overapplication and environmental pollution will be reduced (Economist, 2014b).
- More comprehensive understanding of how chemicals affect all aspects of agriculture, public health, and the environment will contribute to development of more ecologically intense (aware may be a better word) precision farming systems.
- Placement of microscale devices within livestock to monitor and regulate all internal chemical processes or possibly change the operative chemistry.
- Perhaps the most interesting will be research that reveals how evolution has been controlled by chemistry and thereby will give us greater understanding of a plant's chemistry, and genetic potential.

Chemistry has and will continue to contribute to meeting the challenge of feeding all. However, it is important to remember that although present and future food production systems may come closer to meeting the need of all, they will never be able to meet the wants of all.

REFERENCES

Avery, D., 1997. Saving the planet with pesticides and biotechnology and European farm reform. In: British Crop Prot. Conf.–Weeds, pp. 3–18.

David, K., Thompson, P.B. (Eds.), 2008. What Can Nanotechnology Learn from Biotechnology? Social and Ethical Lessons for Nanoscience from the Debate over Agrifood Biotechnology and GMOs. Academic Press, London, UK, p. 342.

Diamond, J., 1987. The worst mistake in the history of the human race. Discover Magazine 18 (5), 64–66.

Dickens, C., 1859. A Tale of Two Cities. Chapman and Hall, London, UK. 295 pp.

Economist, May 24, 2014a. Schumpeter–Digital Disruption on the Farm. p. 64.

Economist, December 6, 2014b. The Robot Overhead. pp. 11–12.

Fox, T., 2013. Global Food Waste Not, Want Not. Institute of Mechanical Engineers, London, U.K. 36 pp.

Gunders, D., 2012. How America Is Losing up to 40% of Its Food from Farm to Fork to Landfill. Natural Resources Defense Council, New York. August IP:12–06B.

Hubert, B., Caron, P., Guyomard, H., 2014. Conclusion. In: Paillard, S., Treyer, S., Dorin, B. (Eds.), Agrimonde–Scenarios and Challenges for Feeding the World in 2050. Springer, London, UK, pp. 197–203. 250 pp.

Keynes, J.M., 1930. Economic possibilities for our grandchildren. In: Essays in Persuasion. W.W. Norton & Co, New York, pp. 358–373. 1963.

Kolbert, E., May 2014. No Time: How Did We Get so Busy. The New Yorker. pp. 70–72, 74–75.

Kunstler, J.H., 2012. Making other arrangements–a wake-up call to a citizenry in the shadow of oil scarcity. In: An Orion Reader. The Orion Society, Great Barrington, MA, pp. 42–52. 107 pp.

Kunzig, R., 2011. Population 7 billion. National Geographic 219 (1), 34, 35, 37, 38, 40, 42–43, 45, 47, 54, 56, 59–64, 67, 68.

Kunzig, R., 2014. A World with 11 Billion People? New Population Projections Shatter Earlier Estimates. National Geographic News. Available only on-line @ http://news.nationalgeographic.com/news/2014/09/140918-population-global-united-nations-2100-boom-africa/ (accessed November 2014).

Le Couteur, P., Burreson, J., 2003. In: Tarcher, Jeremy P. (Ed.), Napoleon's Buttons–17 Molecules that Changed History. Penguin, New York, p. 375.

Singer, P., 2009. The Life You Can Save. Random House, New York. 206 pp.

Suppe, F., 1987. The limited applicability of agricultural research. Agriculture and Human Values 4, 4–14.

Verne, J., 2009. A Journey to the Center of the Earth. Penguin Classics. 252 pp. First published in 1864.

Index

Printed in the United States
By Bookmasters